Analog Circuits and Signal Processing

Series Editors:

Mohammed Ismail, Dublin, USA
Mohamad Sawan, Montreal, Canada

The Analog Circuits and Signal Processing book series, formerly known as the Kluwer International Series in Engineering and Computer Science, is a high level academic and professional series publishing research on the design and applications of analog integrated circuits and signal processing circuits and systems. Typically per year we publish between 5–15 research monographs, professional books, handbooks, edited volumes and textbooks with worldwide distribution to engineers, researchers, educators, and libraries.

The book series promotes and expedites the dissemination of new research results and tutorial views in the analog field. There is an exciting and large volume of research activity in the field worldwide. Researchers are striving to bridge the gap between classical analog work and recent advances in very large scale integration (VLSI) technologies with improved analog capabilities. Analog VLSI has been recognized as a major technology for future information processing. Analog work is showing signs of dramatic changes with emphasis on interdisciplinary research efforts combining device/circuit/technology issues. Consequently, new design concepts, strategies and design tools are being unveiled.

Topics of interest include:

Analog Interface Circuits and Systems;

Data converters;

Active-RC, switched-capacitor and continuous-time integrated filters;

Mixed analog/digital VLSI;

Simulation and modeling, mixed-mode simulation;

Analog nonlinear and computational circuits and signal processing;

Analog Artificial Neural Networks/Artificial Intelligence;

Current-mode Signal Processing; Computer-Aided Design (CAD) tools;

Analog Design in emerging technologies (Scalable CMOS, BiCMOS, GaAs, heterojunction and floating gate technologies, etc.);

Analog Design for Test;

Integrated sensors and actuators; Analog Design Automation/Knowledge-based Systems; Analog VLSI cell libraries; Analog product development; RF Front ends, Wireless communications and Microwave Circuits;

Analog behavioral modeling, Analog HDL.

More information about this series at http://www.springer.com/series/7381

Jiawei Xu • Refet Firat Yazicioglu
Chris Van Hoof • Kofi Makinwa

Low Power Active Electrode ICs for Wearable EEG Acquisition

 Springer

Jiawei Xu
Holst Centre / imec
Eindhoven, The Netherlands

Refet Firat Yazicioglu
Galvani Bioelectronics
Stevenage, United Kingdom

Chris Van Hoof
ESAT-MICAS
KU Leuven / imec
Leuven, Belgium

Kofi Makinwa
Delft University of Technology
Delft, The Netherlands

ISSN 1872-082X ISSN 2197-1854 (electronic)
Analog Circuits and Signal Processing
ISBN 978-3-319-89284-9 ISBN 978-3-319-74863-4 (eBook)
https://doi.org/10.1007/978-3-319-74863-4

This Springer imprint is published by Springer Nature
The registered company is Springer International Publishing AG
The registered company address is: Gewerbestrasse 11, 6330 Cham, Switzerland

Preface and Acknowledgements

This book is the outcome of my PhD research carried out at Holst Centre/imec and Delft University of Technology (TU Delft) between 2010 and 2016. In 2006, Holst Centre lunched the Human++ research program to develop sensors, integrated circuits (ICs), and systems toward personalized wearable healthcare. As a pioneer of biomedical ICs and systems design, imec developed multiple low-power electro-encephalography (EEG) acquisition ICs for ambulatory brain monitoring. My research target was to follow and extend this interesting topic by enabling long-term and continuous EEG recording for improved user comfort while still achieving medical grade signal quality.

One barrier used to hamper convenient and comfortable EEG monitoring is the use of gel (wet) electrodes. As standard electrodes for clinical EEG recording, gel electrodes provide the best signal quality. However, they are not so user-friendly because of several drawbacks, such as skin preparation, electrode replacement, and allergic reaction. An alternative solution is the use of dry (or gel-free) electrodes, which facilitate easy, fast, and comfortable EEG recording at the cost of reduced signal quality. This is because dry electrodes induce very high impedance between skin-electrode interface, resulting in increased noise and interference. Hence, conventional EEG acquisition ICs are not suitable for dry electrodes.

This book describes the application, theory, implementation, and evaluation of active electrode (AE)-based ICs and systems for gel-free EEG acquisition. The AE-based ICs are compatible with dry electrodes because of their robustness to environmental interference and cable motion. To overcome the performance limitations of prior art AE-based systems, such as gain mismatch and low power efficiency, three generations of AEs were implemented with innovative architectures and circuit design techniques. These works show very promising results compared to state-of-the-art wet electrode-based EEG ICs and illustrate structured circuit design methodologies for general biopotential signals acquisition, such as EEG, ECG and EMG.

Undertaking this challenging research has been a rocky road, but definitely not a lonely journey. I would like to thank my colleagues and friends at imec and TU

Delft. As the sources of friendships as well as good collaboration, they have contributed immensely to my personal and professional time in the past 10 years. Special thanks go to Prof. Johan H. Huijsing and Dr. Pieter Harpe for their amazing enthusiasm, patience, and willingness to help me through many discussion and brainstorms.

My sincere thanks go to my PhD supervisors and the co-authors of this book: Prof. Kofi Makinwa, Prof. Chris Van Hoof, and Dr. Firat Yazicioglu. It has been my great fortune and pride to have worked with you. Your contributions of time, ideas, and even criticism make my research so productive and stimulating.

Lastly, my gratitude and love to my parents for always believing in me and supporting me to follow my dreams. And most of all for Liangliang who has been by my side with great patience, understanding, and encouragement.

Eindhoven, The Netherlands   Jiawei Xu
December 2017

Contents

1 Introduction   1
  1.1 Wearable EEG Devices   1
  1.2 Prior-Art EEG Systems   3
  1.3 A Promising Solution: Active Electrodes   5
  1.4 Challenges in Active Electrode Systems   6
  1.5 Book Contributions and Organization   8
  References   9

2 Review of Bio-Amplifier Architectures   11
  2.1 Bio-Amplifier Design Techniques   11
    2.1.1 Chopper Modulation   11
    2.1.2 Impedance Bootstrapping   11
    2.1.3 Offset Compensation   12
    2.1.4 Driven-Right-Leg (DRL)   13
  2.2 Bio-Amplifier Architectures   14
    2.2.1 Analog Buffers   14
    2.2.2 Inverting AC-Coupled Amplifiers   14
    2.2.3 Non-Inverting AC-Coupled Amplifiers   16
    2.2.4 Instrumentation Amplifiers   17
    2.2.5 "Functionally" DC-Coupled Amplifiers   18
    2.2.6 Summary   20
  References   20

3 An Active Electrode Readout Circuit   23
  3.1 IC Architecture Overview   23
  3.2 Active Electrode ASIC   23
    3.2.1 An AC-Coupled Inverting Chopper Amplifier   24
    3.2.2 Digitally Assisted Ripple and Offset Reduction   26
    3.2.3 Input Impedance Boosting   27
    3.2.4 Noise Analysis   31
  3.3 Back-End CMFB IC   32

    3.4 Measurement..   36
         3.4.1 IC Measurement..............................   36
         3.4.2 Cable Motion and Interference...............   38
         3.4.3 Biopotential EEG Measurement................   40
    3.5 Conclusions...   46
    References..   46

4 **An Eight-Channel Active Electrode System**..................   49
    4.1 IC Architecture Overview.............................   49
         4.1.1 EEG and ETI Measurement.....................   49
         4.1.2 A CMFF Technique for CMRR Enhancement........   51
         4.1.3 PWM Communication...........................   53
    4.2 Active Electrode ASIC................................   54
         4.2.1 Instrumentation Amplifier...................   54
         4.2.2 Noise Analysis...............................   55
         4.2.3 Current Source for ETI Measurement...........   57
    4.3 Back-End Analog Signal Processing ASIC..............   57
         4.3.1 Instrumentation Amplifiers..................   57
         4.3.2 Low-Pass Filter and ADC.....................   59
    4.4 Measurement...   59
    4.5 A Four-Channel Wireless EEG Headset................   65
    4.6 Conclusions...   67
    References..   67

5 **Current Noise of Chopper Amplifiers**.......................   69
    5.1 Chopping and Current Noise..........................   69
    5.2 Current Noise Analysis..............................   70
         5.2.1 Charge Injection and Clock Feedthrough.......   70
         5.2.2 Shot Noise from the MOSFETs Channel Charge...   72
         5.2.3 *KT/C* Noise from the Clock Driver..........   72
         5.2.4 Parasitic Switched-Capacitor Resistance.....   73
         5.2.5 Summary.......................................   74
    5.3 Current Noise Measurement...........................   75
         5.3.1 A Conventional Chopper Modulated Amplifier...   75
         5.3.2 Chopper Amplifiers with Capacitive Feedback...   80
    5.4 A Dedicated Noise-Testing Chip......................   87
    5.5 Methods of Reducing Current Noise...................   90
    5.6 Conclusions...   90
    References..   91

6 **A Digital Active Electrode System**.........................   93
    6.1 IC Architecture Overview.............................   93
    6.2 Analog Signal Processing............................   95
         6.2.1 A "Functionally" DC-Coupled Instrumentation
                Amplifier.....................................   95
         6.2.2 Programmable Gain Amplifier..................   98

6.3 Digital Interfaces 100
6.4 CMRR Enhancement 102
6.5 Measurement 104
    6.5.1 Measurement of Performance 104
    6.5.2 Multiparameter ExG Measurement 109
6.6 Conclusions 111
References 113

7 Conclusions 115
7.1 Summary 115
7.2 Future Work 116
References 117

Summary 119

List of Publications 121

Index 123

About the Author

Jiawei Xu received his MSc and PhD degrees in 2006 and 2016, both in Electrical Engineering from Delft University of Technology, The Netherlands. From 2006, he has been working at Holst Centre/imec on low-power sensor interfaces and wearable biomedical ICs. He is currently a senior researcher, leading the R&D of noninvasive brain monitoring ICs. He has developed biosignal acquisition ICs for EEG, ECG, bioimpedance, galvanic skin response (GSR), and near-infrared spectroscopy (NIRS).

Dr. Xu was the recipient of the IEEE Solid-State Circuits Society (SSCS) Predoctoral Achievement Award (2014) and the imec Scientific Excellence Award (2014).

Refet Firat Yazicioglu is the head of Neuromodulation Devices at Galvani Bioelectronics R&D and responsible for the development of implantable devices and creation of new technologies for implantable devices. He received his PhD degree from KU Leuven in Belgium and worked 13 years at imec, Europe's largest independent research center in microelectronics and nanoelectronics. He has developed wearable and implantable medical devices, including wireless cardiac monitoring patches, wearable EEG monitoring headsets, and implantable neural probes for high density recording.

Dr. Yazicioglu has served in the technical program committees of the European Solid-State Circuits Conference (ESSCIRC), the International Solid-State Circuits Conference (ISSCC), and the Biomedical Circuits and Systems Conference (BioCAS). He is Associate Editor for IEEE Transactions on Biomedical Circuits and Systems.

Chris Van Hoof is Director Wearable Health Solutions at imec and imec fellow. Chris leads imec's wearable health R&D across three imec sites (Eindhoven, Leuven, and Gent). Together with his team, he provides solutions for chronic disease patient monitoring and for preventive health through virtual coaching. He is passionate about making things that really work and apart from delivering industry-relevant and fully qualified solutions to customers, his work resulted in four imec startups (three in the healthcare domain).

After receiving a PhD from the University in Leuven in 1992 in collaboration with imec, Chris has held positions as manager and director in diverse fields (sensors, imagers, 3D integration, MEMS, energy harvesting, body-area-networks, biomedical electronics, wearable health). He has published over 600 papers in journals and conference proceedings and has given over 70 invited talks. He is also full professor at the University of Leuven (KU Leuven).

Kofi Makinwa holds degrees from Obafemi Awolowo University, Ile-Ife (BSc, MSc); Philips International Institute, Eindhoven (MEE); and Delft University of Technology, Delft (PhD). From 1989 to 1999, he was a research scientist at Philips Research Laboratories, where he designed sensor systems for interactive displays, and analog front-ends for optical and magnetic recording systems. In 1999 he joined Delft University of Technology, where he is currently an Antoni van Leeuwenhoek Professor of the Faculty of Electrical Engineering, Mathematics and Computer Engineering, and Chair of the Electronic Instrumentation Laboratory.

Dr. Makinwa holds 21 patents, and has authored or coauthored 6 books and over 200 technical papers. He is on the program committee of the European Solid-State Circuits Conference (ESSCIRC) and the workshop on Advances in Analog Circuit Design (AACD). He has also served on the program committees of the International Solid-State Circuits Conference (ISSCC); the International Conference on Solid-State Sensors, Actuators, and Microsystems (Transducers); and the IEEE Sensors

Conference. He was a distinguished lecturer of the IEEE Solid-State Circuits Society (2008–2011) and a guest editor of the Journal of Solid-State Circuits (JSSC). He has given invited talks and tutorials at several international conferences, including ISSCC, ESSCIRC, ASSCC, and the VLSI symposium. At the 60th anniversary of ISSCC, he was recognized as one of its top ten contributing authors.

For his PhD research, Dr. Makinwa was awarded the title of "Simon Stevin Gezel" by the Dutch Technology Foundation (STW). In 2005, he received a VENI grant from the Dutch Scientific Foundation (NWO). He is a corecipient of several best paper awards: from the JSSC (2), ISSCC (4), ESSCIRC (2), and Transducers (1). He is an IEEE Fellow, an alumnus of the Young Academy of the Royal Netherlands Academy of Arts and Sciences (KNAW), and an elected member of the AdCom of the IEEE Solid-State Circuits Society.

Chapter 1
Introduction

1.1 Wearable EEG Devices

In modern clinical practice, scalp EEG measurement is the most important nonin-vasive procedure to measure brain electrical activity and evaluate brain disorders. Electroencephalograms (EEGs) represent the brain's spontaneous electrical activi-ties by measuring scalp potentials over multiple areas of the brain (Fig. 1.1) [1], so the strength and distribution of such potentials reflects the average intensity and position of a group of underlying neurons. As a noninvasive method, EEGs play a vital role in a wide range of clinical diagnosis, such as epileptic seizures, Alzheimer's disease, and sleep disorders [2]. Furthermore, EEGs are also finding increasing popularity in nonclinical neuroscience and cognitive research [3]. Typical applications include brain-computer interfaces (BCI), neurofeedback, or brain func-tion training.

During the last decade, there is a growing need toward continuous monitoring of brain activities in remote patient monitoring, health, and wellness management. These come from the increased prevalence of chronic diseases and the need to decrease the length of hospital stays [4]. The huge market demand, together with the advances in electronic manufacturing techniques, has accelerated the evolution of power-efficient and miniaturized wearable sensors for biomedical applications (Fig. 1.2), with long-term monitoring and user-friendliness being the key drivers.

Although the first human EEG recording device (Fig. 1.3a) was invented in 1924, a personalized EEG recording system for residential monitoring was not available until the 1970s [6]. Later, ambulatory EEG systems (Fig. 1.3b) and portable EEG devices (Fig. 1.3c) in principle gave users sufficient mobility during the recording. However, these devices are still bulky and power hungry and are therefore unsuitable for long term and continuous EEG recording.

Most recent advances in biomedical techniques, sensors, integrated circuits (ICs), batteries, and wireless communication have sped up the development of real "wear-able" EEG monitors. For example, a miniature, lightweight, and battery-powered

© Springer International Publishing AG 2018
J. Xu et al., *Low Power Active Electrode ICs for Wearable EEG Acquisition*, Analog
Circuits and Signal Processing, https://doi.org/10.1007/978-3-319-74863-4_1

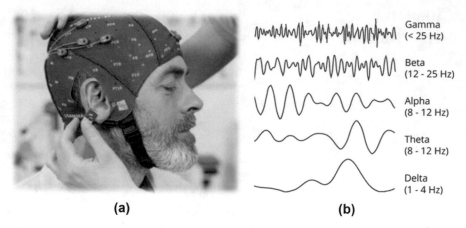

(a)   (b)

Fig. 1.1 (a) Wearable EEG measurement. (b) Typical electrical signals from the brain

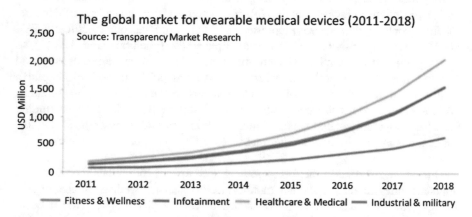

Fig. 1.2 Market growth trends of wearable technology [5]. The global market for wearable medical devices was valued at USD 750 million in 2012 and is expected to reach a value of USD 5.8 billion in 2018, growing at a compound annual growth rate (CAGR) of 40.8% from 2012 to 2018

wireless EEG recording unit (Fig. 1.3d) can be implemented inside various easy-to-use form factors [10–12], such as EEG caps, headsets, or helmets. These EEG units collect raw data of brain activities during a user's daily routine, which can then be used to extract biomarkers and to determine personal trends for emotion, behavior, disease management, and wellness applications.

This book presents new generations of energy-efficient EEG signal acquisition ICs, which are typically the core of an EEG monitor and dominate its overall performance. The electronic design methodologies and detailed implementation of the ICs toward wearable applications are discussed.

(a) First EEG recording

(b) Conventional EEG system

(c) Portable EEG system

(d) Wearable EEG system

Fig. 1.3 Evolutions in EEG readout systems: (**a**) the first recording of human EEGs [7] (Hans Berger, 1924), (**b**) a 192-channel EEG system [8] (Nihon Kohden, 1999), (**c**) a portable EEG-based BCI system [9] (g.tec, 2003), and (**d**) an 8-channel wireless EEG headset [10] (imec/Holst Centre, 2013)

1.2 Prior-Art EEG Systems

As a standard practice, a single-channel EEG acquisition instrument contains three electrodes, three lead wires, and a differential instrumentation amplifier (IA) (Fig. 1.4). The instrument measures the difference in voltage between one electrode and the reference electrode. Both electrodes convert ionic current into electric current. The EEG potential represents voltage fluctuations resulting from ionic current within brain neurons. Via two lead wires, an IA amplifies the differential EEG potential between these two electrodes. A third electrode, namely, the bias electrode or ground electrode, helps keep the body's DC voltage level in-line with the readout circuits to properly amplify the EEG signal. Without the bias electrode connected to the body, the electrode potentials may drift and, eventually, saturate the IA's input.

In the electrical domain, the electrode-tissue interface can be modeled as a complex impedance in series with a DC voltage source, which represents the polarization voltage between skin and electrode (Fig. 1.5).

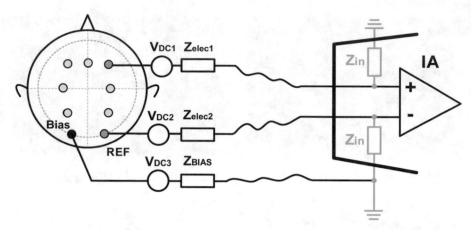

Fig. 1.4 Acquiring an EEG signal through three (passive) electrodes and a differential instrumentation amplifier

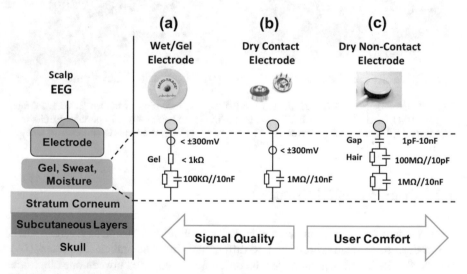

Fig. 1.5 Equivalent electrical model of the electrode-tissue interface [13]

The biggest challenge facing designers of wearable EEG systems is achieving improved user comfort, long-term monitoring capability with medical-grade signal quality. Unfortunately, prior-art EEG systems rarely meet all these requirements.

One major drawback of prior-art EEG systems is their dependence on gel or wet electrodes. Conductive gel reduces skin-electrode impedance and the associated artifacts induced by cable motion. Therefore, wet electrodes are extensively used in clinical practice. However, wet electrodes require skin preparation and professional personnel to place them properly. In addition, the gel can dry out and therefore

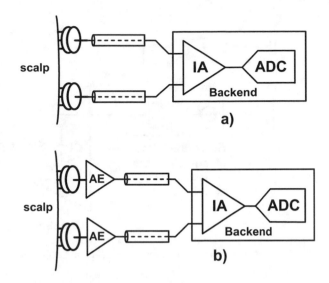

Fig. 1.6 Illustration of EEG readout circuits: (**a**) conventional solution based on an IA and (**b**) proposed solution based on active electrodes

requires frequent replacement of electrodes to maintain signal integrity. These drawbacks limit the use of wet electrodes in wearable EEG applications.

Dry electrodes, on the other hand, facilitate long-term EEG recording as well as greater user comfort. However, this comes at the expense of reduced signal quality due to the larger skin-electrode impedance (Fig. 1.5). This complex impedance can be as high as a few MΩ at 50/60 Hz [14, 15], and it significantly increases noise and interference pickup from the environment. Dry electrodes thus need to be buffered or shielded in order to approach the performance of wet electrodes [16]. Directly connecting dry electrodes to an EEG amplifier via light-weight non-shielded cables will not ensure good signal quality. Hence, conventional EEG recording systems (Fig. 1.6a), based on passive electrodes connected to a differential biopotential amplifier through long cables, are ill-suited for the use with dry electrodes and wearable devices.

1.3   A Promising Solution: Active Electrodes

Active electrodes (AEs) with co-integrated amplifiers solve this incompatibility problem (Fig. 1.6b). The close proximity of the electrodes to the amplifier reduces interference pickup, while the amplifier's low output impedance improves signal robustness to cable motion [16]. Moreover, the signal quality of dry-electrode EEG recording can be maintained without using conventional shielded cables, which is an attractive feature for compact wearable devices.

Early AEs consisted of simple analog buffers (i.e., voltage followers). Improved buffer-based AEs achieved higher input impedance [17] or required fewer cables [18, 19]. The main limitation of this classic AE topology is its power inefficiency, as

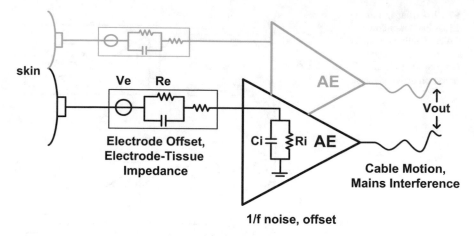

Fig. 1.7 Aggressors in the skin-electrode interface and active electrodes

an analog buffer only performs impedance conversion without providing any voltage gain. The succeeding backend readout circuits still need to meet the same specifications of noise and electrode offset tolerance, leading to additional power consumption [20].

In contrast, this book presents three generations of amplifier-based AEs implemented with power-efficient instrumentation amplifiers (IA). Although interfacing with dry electrodes and reducing the overall system power are the primary motivations for selecting the amplifier-based AE architectures, the proposed amplifier-based AE systems also aim to achieve performance that is comparable with that of medical grade systems.

1.4   Challenges in Active Electrode Systems

AEs constitute the first stage of a wearable EEG system, and thus determine its overall performance. Therefore, AEs should cope with the same challenges as conventional differential amplifiers (Fig. 1.7):

- Amplifying low-frequency low-amplitude EEG signal
- Interfacing with high impedance between skin and electrode
- Tolerating large electrode polarization voltages
- Suppressing environmental artifacts
- Minimizing system volume and power dissipation

In addition, the AEs also need to solve various practical challenges:

- Reducing the component mismatch between AEs
- Minimizing the number of connecting wires

Table 1.1 Medical standards and proposed specification

	IEC60601-2-26[a][21]	IFCN[b][22]	Proposed specifications (wearable EEG)
Input voltage range	$1\ mV_{pp}$	–	$>1\ mV_{pp}$
Input referred noise (per channel)	$6\ \mu V_{pp}$	$1.5\ \mu V_{pp}$ $0.5\ \mu V_{rms}$ (0.5–100 Hz)	$1\ \mu V_{rms}$ (0.5–100 Hz)
HPF cutoff frequency	< 0.5 Hz	< 0.16 Hz	< 0.5 Hz
Electrode offset tolerance	±300 mV	–	±300 mV
Input impedance (at 50/60 Hz)	–	>100 MΩ	> 100 MΩ
CMRR (at 50/60 Hz)	–	110 dB	>80 dB
Power consumption (per channel)	–	–	<100 μW
Applications	Wet electrodes Clinical	Wet electrodes Clinical	Dry electrodes BCI, wellness

[a]IEC60601 is a series of technical standards for the safety and effectiveness of medical electrical equipment published by the International Electrotechnical Commission
[b]IFCN stands for International Federation of Clinical Neurophysiology

The sections below discuss these challenges in detail. In general, the major specifications of bio-amplifiers for clinical EEG applications are defined and driven by medical standards (Table 1.1). For wearable EEG devices with dry electrodes, the electrode-tissue impedance (ETI) can get quite large. As a result, the required specifications of input impedance, electrode offset tolerance, and power dissipation are even tighter, while the CMRR and noise specifications of the bio-amplifier itself can be slightly relaxed.

EEG reflects the summation of the electrical activity of thousands or millions of neurons under the scalp. A typical adult EEG has an amplitude of 10–100 μV when measured on the scalp, and it increases to 10–20 mV when measured by subdural electrodes [23]. Most of the cerebral signal observed in a scalp EEG falls in the range of 1–30 Hz (activity below or above this range is likely to be caused by artifacts, under standard clinical recording conditions). The EEG rhythmic activity is divided into frequency bands, such as delta (<4 Hz), alpha (8–15 Hz), and beta (16–31 Hz), which are used to detect various physiological behaviors. To amplify such low-frequency and low-amplitude potentials, an EEG IA should have the maximum input referred noise of $6\ \mu V_{pp}$ [24].

The use of dry electrodes comes with large and variable skin-electrode imped-ance, as well as large electrode polarization voltages. AEs must then have very high input impedances (>100 MΩ at 50/60 Hz) to minimize signal attenuation. Electrode polarization voltage, or half-cell potential, develops across the electrolyte-electrode interface due to an uneven distribution of anions and cations [15]. This offset voltage per electrode can be as large as a few hundreds of mV and may saturate the IA. Thus, the IA should be able to tolerate at least 300 mV electrode DC offset [24] while still maintaining its noise performance.

Mains interference can be picked up from the environment during EEG acquisition, because a high-impedance (dry) electrode behaves like an antenna. Although this issue can be mitigated by using AE architectures with low output impedance, AE mismatch can still convert any common-mode (CM) interference and motion artifact into a differential signal. Such signals can be larger than the μV EEG signals, thus reducing the dynamic range of the AE and requiring complex post-filtering. This can be avoided by designing AE pairs with a high common-mode rejection ratio (CMRR).

A miniaturized multichannel AE system requires a minimal number of wires connected to a backend circuit. This reduces the overall cabling, which is especially necessary when tens of AEs are used for multichannel acquisition or when additional wires are needed for multiparameter measurement [25, 26]. A nice example of multiparameter measurement involves recording EEG and electrode-tissue impedance (ETI) simultaneously; the ETI provides information that can be used for impedance based motion artifact reduction or simple lead-on/off detection.

Finally, a battery-powered AE system should consume ultralow power to maximize its operating time. For example, to realize 24-h continuous operation with a 3.6 V coin cell battery [27], an AE system, including multiple AEs and a backend readout circuit, must consume less than 5 mA. Although a battery with more energy capacity can be used, its size and weight will be a major determinant of the system's form factor.

In summary, an ideal AE system should balance the tradeoff among different parameters to maximize overall performance, even in the presence of dry electrodes and the aforementioned aggressors.

1.5   Book Contributions and Organization

A complete EEG signal processing chain for emerging wearable applications usually contains these major building blocks: analog front-ends (AFE), digital signal processing (DSP), a wireless transmitter, and power management units (PMU). This book focuses on the design of EEG AFE, with a special emphasis on instrumentation amplifier (IA) architecture and design for AEs.

The main contributions of this work include the following:

- Analysis of capacitively-coupled AE architectures. Three types of chopper amplifiers are used as AEs that balance the tradeoff between noise, electrode offset tolerance, input impedance, and power consumption. The overall performance of these proposed AEs is competitive with state-of-the-art biopotential IAs through the use of various circuit design techniques. These techniques include positive feedback, which increases AE's input impedance by a factor of 5–10 (Chap. 3); digitally assisted calibration, which reduces AEs' non-idealities (ripple and offset) by a factor of 10 (Chap. 3); and a functionally DC-coupled AE, which enables an electrode offset tolerance of up to ±350 mV while consuming very low power (Chap. 6).

- Development of CMRR boosting techniques that overcome the CMRR limitations imposed by AE gain mismatch. These techniques include a common-mode feedback (CMFB) circuit that processes the AEs' outputs and feeds their common-mode signal back to each AE (Chap. 3), a power-efficient common-mode feedforward (CMFF) technique that creates a voltage averaging node to reduce the AEs' common-mode current (Chap. 4), and a generic CMFF approach that utilizes an analog buffer to drive the AE's negative input, thus cancelling input CM interference before amplification (Chap. 6).
- Investigation of current noise, which can be a significant noise contributor of chopper amplifiers. Chopping was observed to cause excess current noise, which, at high-impedance nodes, is converted into voltage noise with a slope of $1/f^2$. The origin of this noise is hypothesized to be the charge injection and clock feedthrough of the input chopper. This current noise theory has been analyzed and experimentally verified (Chap. 5).
- Design of a single-chip digital active electrode (DAE) architecture, which combines an IA, an ADC and an I^2C interface for on-chip analog signal processing and digitization (Chap. 6). This DAE architecture enables a daisy chain connection of all DAEs and a generic µC on a two-wire I^2C bus, significantly reducing system complexity and cost.

The book is organized as follows: Chapter 1 introduces the basics of scalp EEG measurement, dry-electrode interfaces, AEs, and the associated design challenges. Chapter 2 reviews the AE and IA architectures and compares their performance tradeoffs. Chapter 3 presents the use of an AC-coupled inverting IA as an AE. - Chapter 4 presents a complete 8-channel AE-based EEG recording system, including both front-end AEs and a back-end analog signal processor (ASP). Chapter 5 describes an experimental investigation of current noise in chopper amplifiers. Chapter 6 presents a digital active electrode (DAE), with built-in IAs, an ADC, and a digital interface on a single chip. Chapter 7 concludes this book and gives directions for future work.

References

1. G.tec Gamma. [online] available at: http://www.gtec.at/Products/Electrodes-and-Sensors/g. GAMMAsys-Specs-Features
2. O.V. Lounasmaa et al., Information processing in the human brain – Magnetoencephalographic approach. Proc. Natl. Acad. Sci. U. S. A. **93**, 8809–8815 (1996)
3. EEGinfo. What is neurofeedback? [online] available at: https://www.eeginfo.com/what-is-neurofeedback.jsp
4. D.E. Bloom et al., *The global economic burden of noncommunicable diseases* (World Economic Forum, Geneva, 2011)
5. Transparency Market Research. [online] available at: http://www.transparencymarketresearch.com/wearable-medical-devices.html
6. G.B. Marson, J.B. McKinnon, A miniature tape recorder for many applications. Control. Instrum. **4**, 46–47 (1972)

7. IBVA. [online] available at: http://www.ibva.co.uk/eeg.htm
8. Neurofax EEG-1100 System (1999), Nihon Kohden
9. g.tec IntendiX. [online] available at: http://www.cortechsolutions.com/Products/DA/DA-IX
10. imec EEG headset (2012). [online] available at: http://www2.imec.be/be_en/press/imec-news/imeceeg2012.html
11. Cognionics, 72-Channel Dry EEG Headset System. [online] available at: http://www.cognionics.com
12. Emotiv. [online] available at: https://emotiv.com/
13. Y.M. Chi, T.-P. Jung, G. Cauwenberghs, Dry-contact and noncontact biopotential electrodes: Methodological review. IEEE Rev. Biomed. Eng. **3**, 106–119 (2010)
14. A. Searle, L. Kirkup, A direct comparison of wet, dry and insulating bioelectric recording electrodes. Physiol. Meas. **21**(2), 271 (2000)
15. S. Lee, J. Kruse, Biopotential electrode sensors in ECG/EEG/EMG systems. Analog Devices **200**, 1–2 (2008)
16. A.C. Metting-van Rijn et al., High-quality recording of bioelectric events. Part 2. Low-noise, low-power multichannel amplifier design. Med. Biol. Eng. Comput. **29**(4), 433–440 (1991)
17. C.J. Harland et al., Electric potential probes – New directions in the remote sensing of the human body. Meas. Sci. Technol **13**, 163 (2002)
18. T. Degen et al., Low-noise two-wired buffer electrodes for bioelectric amplifiers. IEEE Trans. Biomed. Eng. **54**, 1328–1332 (2007)
19. F.Z. Padmadinata et al., Microelectronic skin electrode. Sens. Actuators B Chem. **1**(1–6), 491–494 (1990)
20. J. Xu, R. F. Yazicioglu, P. Harpe, K. A. A. Makinwa, C. Van Hoof, A 160µW 8-channel active electrode system for EEG monitoring, *Digest of ISSCC*, (Feb. 2011), pp. 300–302
21. IEC60601-2-26, Medical electrical equipment – Part 2-26: Particular requirements for the basic safety and essential performance of electroencephalographs, (2012)
22. M.R. Nuwer et al., IFCN standards for digital recording of clinical EEG. Electroencephalogr. Clin. Neurophysiol. **106**(3), 259–261 (1998)
23. H. Aurlien et al., EEG background activity described by a largecomputerized database. Clin. Neurophysiol. **115**(3), 665–673 (2004)
24. IEC60601-2-26, 3rd Edition, Medical electrical equipment – Part 2-26: Particular requirements for basic safety and essential performance of electroencephalographs, (2012)
25. S. Kim et al., A 2.4µA continuous-time electrode-skin impedance measurement circuit for motion artifact monitoring in ECG acquisition systems, *Digest of Symp. VLSI Circuits*, (June 2010), pp. 219–220
26. M. Guermandi, R. Cardu, E. Franchi, R. Guerrieri, Active electrode IC combining EEG, electrical impedance tomography, continuous contact impedance measurement and power supply on a single wire, *Digest of ESSCIRC*, (Sept. 2011), pp. 335–338
27. Datasheet LIR2450 – Multicomp, Coin Cell, Lithium, 120mAh, 3.6V. [online] available at: https://www.powerstream.com/p/Lir2450.pdf

Chapter 2
Review of Bio-Amplifier Architectures

2.1 Bio-Amplifier Design Techniques

2.1.1 Chopper Modulation

$1/f$ noise, or flicker noise, is usually the dominant voltage noise source of a bio-amplifier, because the bandwidth of $1/f$ noise is typically in the order of a few kHz, which is far beyond the EEG signal bandwidth of 100 Hz. $1/f$ noise can be reduced by enlarging the size of input transistors. However, using extremely large input transistors not only increases the chip area but also induces significant parasitic capacitance, causing concerns for reduced input impedance and CMRR.

Chopper modulation [1] is a widely-used technique for reducing an IA's low-frequency noise and offset without disturbing the continuous-time operation. In addition, by periodically swapping its inputs, chopping increases an IA's CMRR by averaging its gain mismatch. The operating principle of chopper modulation is shown in Fig. 2.1, where a low-frequency input signal is up-modulated to a chopping frequency (f_c) by a square-wave modulator, and then this signal is amplified by an IA and demodulated back to original baseband by another square wave modulator. On the other hand, the intrinsic offset and $1/f$ noise (below the chopping frequency) of the IA are up-modulated to f_c by the second square wave modulator. These residual signals at f_c, known as ripple, can be filtered by a low-pass filter (LPF) or otherwise suppressed by a ripple reduction loop [2].

2.1.2 Impedance Bootstrapping

AEs require high input impedance to minimize voltage division via skin-electrode impedance, especially in the case of a dry-electrode interface. Impedance bootstrapping has been used to improve an IA's input impedance in various ways,

© Springer International Publishing AG 2018
J. Xu et al., *Low Power Active Electrode ICs for Wearable EEG Acquisition*, Analog
Circuits and Signal Processing, https://doi.org/10.1007/978-3-319-74863-4_2

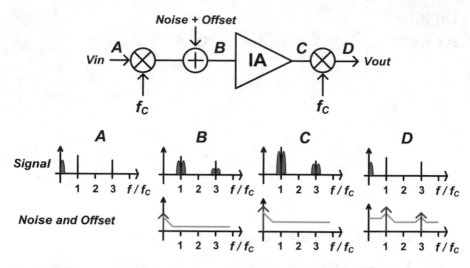

Fig. 2.1 Chopper modulation technique to reduce IA's offset and $1/f$ noise

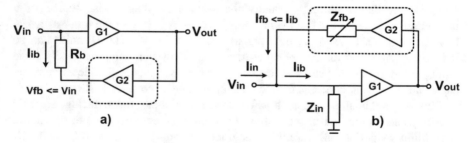

Fig. 2.2 Impedance boosting techniques: (**a**) voltage feedback-based, (**b**) current feedback-based

where a proper positive feedback is often the fundamental element. In [3, 4], the IA's output is fed back to bootstrap its input lead bias resistor (Fig. 2.2a), leading to very high input impedance suitable for noncontact EEG sensing. In [5], the input bias current of the IA is partially provided by a positive feedback loop (Fig. 2.2b), effectively increasing the IA's input impedance. In both cases, the input impedance can be bootstrapped to be infinitely large. Nevertheless, the amount of the positive feedback, either current or voltage, must be carefully controlled to maintain loop stability.

2.1.3   Offset Compensation

Electrode offset, up to a few hundreds of mV, can easily saturate an IA and therefore must be rejected or compensated. AC-coupling via RC components is the most obvious way of electrode offset rejection, as it ensures a rail-to-rail electrode offset

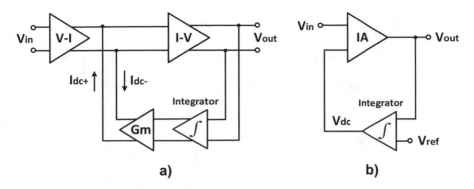

Fig. 2.3 Electrode offset compensation techniques: (**a**) current feedback-based, (**b**) voltage feedback-based

tolerance without consuming any power. However, to eliminate the use of large passive components for better area efficiency, or to further compensate any residual offset, a DC servo loop (DSL) is usually needed.

A DSL is a very effective and probably the only option for electrode offset compensation when passive AC coupling is not available. A DSL based on the negative feedback works as follows: the output offset is tracked and fed back to the input amplifier via current feedback (Fig. 2.3a) [6] or via voltage feedback (Fig. 2.3b) [7]. Both can compensate a certain amount of electrode offset, from a few tens of mV to several hundreds of mV.

2.1.4 Driven-Right-Leg (DRL)

There are two mechanisms that limit the practical CMRR of an EEG acquisition system: mismatch of electrode-tissue impedance (ETI) and gain mismatch of the AEs. The former can be mitigated by maximizing the AE's input impedance, while the latter can be reduced by chopping. Unfortunately, chopping between two AEs is not practical for an AE-based system, where the AEs are mounted on separate electrodes and are placed far from each other. Thus, the component mismatch of the AEs usually results in a low CMRR (<60 dB).

The most well-known circuit for CMRR enhancement is the driven-right-leg (DRL) circuit (Fig. 2.4) [8], where the common-mode (CM) input signal is tracked and fed back to the subject through a third electrode, i.e., the bias electrode. Since the electrode-tissue impedance (Z_e and Z_{rl}) are also in the feedback loop, the DRL technique improves CMRR by reducing the common-mode impedance to the IA, resulting in less pickup of CM interferences from the human body. However, large (external) capacitors (a few nF) and current limiting resistors (a few 100 kΩ) are required to make the loop stable. When dry electrodes are used, it becomes difficult to ensure loop stability over a wide impedance range (100 kΩ–10 MΩ), when both electrode offset and electrode impedance mismatch exist.

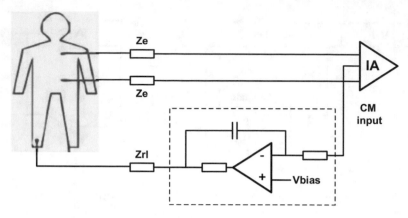

Fig. 2.4 Driven-right-leg (DRL) circuit for CMRR enhancement

2.2 Bio-Amplifier Architectures

2.2.1 Analog Buffers

Most AEs have been simple analog buffers. This confers advantages in terms of large input dynamic range, low output impedance, and low gain variation. Without any added functionality, a buffer requires only a 3-wire connection (V_{dd}, V_{ss}, and V_{out}) to the back-end electronics. Several variants have been published with even fewer wires. In [9], the buffer's analog output is shared with the negative supply voltage of the buffer in a single wire through a current driver, at the cost of less input dynamic range (Fig. 2.5a). Similarly, in [10], the analog output is shared with the positive supply voltage (V_{dd}) of the buffer; however, this requires higher supply voltage and power dissipation (Fig. 2.5b).

A major drawback of buffer-based AE systems is their power efficiency: the buffer requires significant power to meet a low noise specification. However, the buffer only performs impedance conversion without providing any voltage gain or rejecting electrode offset. The subsequent back-end circuit still needs to tackle the same challenges of low noise and large DC tolerance, leading to additional power consumption. A detailed power comparison of AEs, implemented with buffers or amplifiers, will be presented in Chap. 3.

2.2.2 Inverting AC-Coupled Amplifiers

An inverting amplifier with resistive feedback (Fig. 2.6a) is rarely used as a bio-amplifier because the input resistors generate noise and determine IA's input impedance. AC-coupled inverting amplifiers with capacitive feedback (Fig. 2.6b) [11] have been widely used for wearable and implantable medical devices [12, 13]

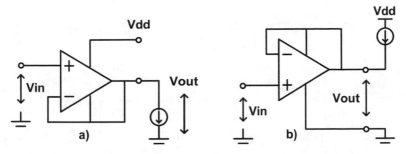

Fig. 2.5 IC techniques to reduce the number of wires of an AE: (**a**) analog output shared with the negative supply voltage of the buffer, (**b**) analog output shared with the positive supply voltage of the buffer

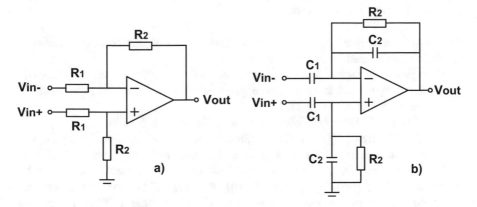

Fig. 2.6 Inverting amplifiers: (**a**) resistive feedback, (**b**) capacitive feedback

because of their rail-to-rail offset rejection capability, area efficiency, and low power consumption. The input coupling capacitor C_1 rejects any electrode offset from the leads. Resistors R_2 can be implemented with MOSFET resistors [11], resulting in resistances of tens of GΩ. This feature makes such IAs easily achieve low cutoff frequencies (< 0.5 Hz) with small on-chip capacitors, in the order of several pF.

The power efficiency of a bio-amplifier can be quantified by the noise efficiency factor (NEF) [14], which represents an IA's noise and power tradeoff in a certain bandwidth.

$$ \text{NEF} = V_{\text{ni,rms}} \sqrt{\frac{2I_{\text{tot}}}{\pi \cdot U_T \cdot 4kT \cdot BW}} \qquad (2.1) $$

where $V_{\text{ni,rms}}$ is the input-referred root mean square (rms) noise voltage, I_{tot} is the total supply current, U_T is the thermal voltage kT/q, and BW is the IA's (-3 dB) bandwidth. State-of-the-art IA with capacitive feedback achieves an NEF of 1.74 [15] by combining a low supply voltage with current reuse techniques. It exploits the

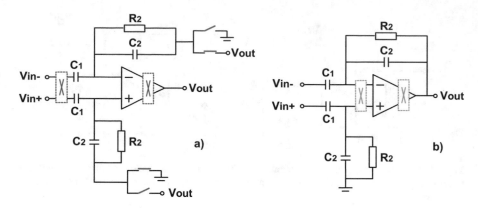

Fig. 2.7 AC-coupled inverting amplifier with alternative chopping schemes: (**a**) before the input capacitor, (**b**) after the input capacitor

fact that the amplifier's input is at virtual ground, and so the core amplifier only needs to have a small input dynamic range.

Chopping can further reduce the $1/f$ noise of this type of IAs. As shown in Fig. 2.7, chopper modulation can be applied at location (a) or (b) to mitigate $1/f$ noise and further improve the NEF.

The IAs in [5, 16] apply input chopper modulation before the input capacitor (Fig. 2.7a) to mitigate $1/f$ noise. One major drawback of this chopper IA topology is its limited tolerance to electrode offset, because it is basically a high-gain DC-coupled amplifier. Although the input DC signal can be partially cancelled by a DC servo loop (DSL) (Fig. 2.8) [5, 16], the tradeoff between the amount of feedback current and the input noise still limits the maximum DC tolerance to a few tens of mV. Furthermore, input impedance of this chopper IA is limited by the switched capacitor impedance associated with its input capacitors. To overcome these issues, an alternative chopping approach places the input chopper inside the capacitive feedback loop (Chap. 3), i.e., at the virtual ground (Fig. 2.7b) [17]. This architecture retains the benefits of a non-chopped capacitive feedback IA, in terms of high input impedance, large electrode offset tolerance, and low power, while mitigating $1/f$ noise through chopping. In addition, an impedance boosting loop, a ripple reduction loop, and an offset calibration loop can be added for even better performance (Chap. 3). A single-ended version of such IAs can also be used as an AE [18].

2.2.3   Non-Inverting AC-Coupled Amplifiers

A non-inverting IA (Fig. 2.9a) has a single-ended input and, since its input impedance is determined by parasitic capacitance, a higher input impedance than an inverting IA. AEs utilizing resistor feedback were published in [19, 20]. However, this is not an area-efficient solution because it requires large and accurate resistors.

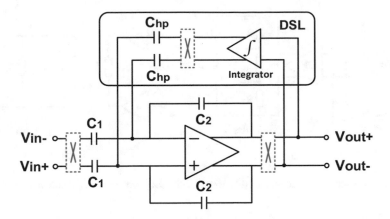

Fig. 2.8 Capacitively-coupled chopper amplifier with a DC servo loop (DSL)

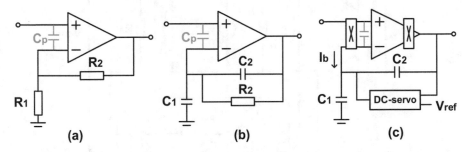

Fig. 2.9 Non-inverting amplifiers: (**a**) resistive feedback IA, (**b**) capacitive feedback IA, (**c**) capacitive feedback IA with a DC servo loop

Moreover, these resistors also increase the input noise. An alternative solution is a capacitive feedback network (Fig. 2.9b) [21, 22], which improves the tradeoff between noise and area efficiency. Moreover, the non-inverting capacitive IA has a DC gain of 1 and so can accommodate relatively large electrode offsets. However, when chopping is utilized, the increased input bias current due to charge injection may create a significant offset voltage via the feedback impedance [17]. Hence, a non-inverting chopper IA usually incorporates a DSL (Fig. 2.9c) to compensate electrode offset [21, 22] (Chap. 4).

2.2.4 Instrumentation Amplifiers

Instrumentation amplifiers (Fig. 2.10) are also widely used in biopotential signal measurements because of their high input impedance. However, a DC-coupled IA [23, 24] has limited electrode offset tolerance. Hence, conventional DC-coupled IAs are not directly applicable to dry-electrode EEG measurement. A DC-coupled

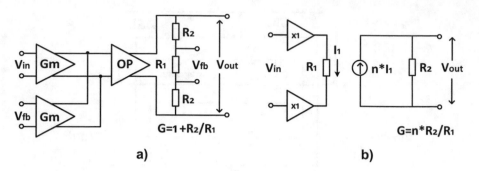

Fig. 2.10 Instrumentation amplifiers: (**a**) current feedback IA architecture, (**b**) current balancing IA architecture

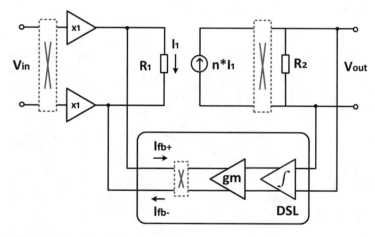

Fig. 2.11 Current-balancing instrumentation amplifier with a DSL

current-balancing IA equipped with a DSL [6] solves this problem (Fig. 2.11) by effectively creating an AC-coupled IA. The IA's noise is further improved through chopping. However, this IA still suffers from limited electrode offset tolerance of a few tens of mV, because the DSL is implemented as a voltage-to-current feedback loop, where a significant amount of feedback current will be required to compensate a large electrode offset.

2.2.5 "Functionally" DC-Coupled Amplifiers

An AC-coupled IA achieves large electrode offset tolerance, but this comes at the cost of filtering out DC and low frequency signals, which may contain useful information, such as low frequency surface potentials [25]. In contrast, a DC-coupled IA preserves such information, but its voltage gain is constrained by

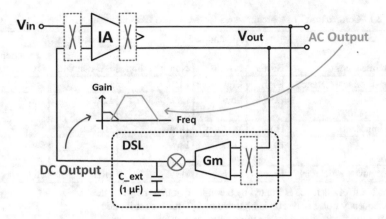

Fig. 2.12 "Functionally" DC-coupled IA

Fig. 2.13 IA with a digitally-assisted offset compensation

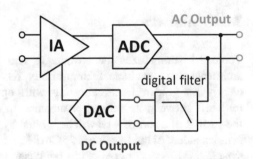

electrode offset and supply voltage and will typically be quite low (<10). As a result, a wide dynamic range, IA typically requires a high-resolution ADC (>16 bit). This, in turn, significantly increases the system's power dissipation, especially when multichannel (>24) EEG acquisition is required, because each channel needs a power-hungry ADC [26].

A "functionally" DC-coupled IA (Fig. 2.12) [7] combines the merits of an AC-coupled IA and a DC-coupled IA, i.e., compensating for large electrode offsets with low power while still being DC-coupled. This is realized by using a voltage-to-voltage feedback (Fig. 2.3b) instead of a voltage-to-current feedback (Fig. 2.3a), which suffers from the tradeoff between electrode tolerance, noise, and power consumption [16]. Although the DSL utilizes an external capacitor to achieve a low cutoff frequency, the "functionally" DC-coupled IA (Chap. 6) can cope with electrode offset of a few hundred mVs. This IA retains the same transfer function as a standard DC-coupled IA, except for the DC signal not being amplified.

This architecture is also applicable to a differential EEG amplifier [27], and the DSL can be implemented in a digitally-assisted manner [28]. Low-pass filtering in the digital domain has the advantage of power and area efficiency. However, since the electrode offset is fed back to the IA through a DAC, quantization noise of the DAC must be reduced. The DC and extremely low frequency signals of the IA are directly available at the DAC's input (Fig. 2.13).

Table 2.1 Comparison of IA architectures for AE-based EEG acquisition

		AC-coupled IA		DC-coupled IA	
AE architectures	Buffer	Inverting	Non-inverting	DC-coupled	"Functionally" DC-coupled
Electrode offset tolerance	High	High	Medium[a]	Low	High
Noise (with chopping)	Low	Low[b]	Low[b]	Low	Low
Input impedance	High	Medium[c]	High	High	High
CMRR[d] (of two IAs)	High	Low	Low	Low	Low
System power efficiency	Low	High	High	High	High

[a]Electrode offset tolerance is limited by the input dynamic range of the IA
[b]Low-frequency noise (<10 Hz) is high due to $1/f^2$ noise contribution
[c]Input impedance is limited by the input coupling capacitors
[d]Without CMRR enhancement techniques

2.2.6   Summary

Table 2.1 summarizes the advantages and limitations of different bio-amplifier architectures to evaluate their usability for wearable EEG acquisition. There is clearly no golden IA architecture with optimum performance because of the tradeoffs between its various parameters. In addition, when IAs are used as AEs, the CMRR of a pair of AEs will be limited by the gain mismatch of their IAs, which is independent of the IAs' intrinsic CMRR, and therefore must be compensated at the system level. A major goal of this book is to explore the circuit design techniques to maximize the IAs' overall performance, at both circuit level and system level, to make them suitable for AE-based EEG acquisition.

References

1. C.C. Enz, G.C. Temes, Circuit techniques for reducing the effects of op-amp imperfections: Autozeroing, correlated double sampling, and chopper stabilization. Proc. IEEE **84**, 1584–1614 (1996)
2. R. Wu, K.A.A. Makinwa, J.H. Huijsing, A chopper current-feedback instrumentation amplifier with a 1mHz 1/f noise corner and an AC-coupled ripple reduction loop. IEEE J. Solid State Circuits **44**(12), 3232–3243 (2009)
3. C.J. Harland, T.D. Clark, et al., Electric potential probes – New directions in the remote sensing of the human body. Meas. Sci. Technol **13**, 163 (2002)
4. Y.M. Chi, C. Maier, G. Cauwenberghs, Ultra-high input impedance, low noise integrated amplifier for noncontact biopotential sensing. IEEE J. Emerging Sel. Top. Circuits Syst. **1**(4), 526–535 (2011)
5. Q. Fan et al., A 1.8μW 60nV/√Hz capacitively-coupled chopper instrumentation amplifier in 65nm CMOS for wireless sensor nodes. IEEE J. Solid State Circuits **46**(7), 1534–1543 (2011)
6. R.F. Yazicioglu, P. Merken, et al., A 60μW 60 nV/√Hz readout front-end for portable biopotential acquisition systems. IEEE J. Solid State Circuits **42**(5), 1100–1110 (2007)

7. J. Xu, B. Büsze et al., A 60nV/sqrt(Hz) 15-channel digital active electrode system for portable biopotential acquisition, *Digest of ISSCC*, (Feb. 2014), pp. 424–425

8. B.B. Winter, J.G. Webster, Driven-Right-Leg circuit design. IEEE Trans. Biomed. Eng. **30**(1), 62–66 (1983)

9. F.Z. Padmadinata et al., Microelectronic skin electrode. Sens. Actuators B **1**(1–6), 491–494 (1990)

10. T. Degen et al., Low-noise two-wired buffer electrodes for bioelectric amplifiers. IEEE Trans. Biomed. Eng. **54**, 1328–1332 (2007)

11. R.R. Harrison, C. Charles, A low-power low-noise CMOS amplifier for neural recording applications. IEEE J. Solid State Circuits **38**(6), 958–965 (2003)

12. F. Shahrokhi, K. Abdelhalim, D. Serletis, P. Carlen, R. Genov, The 128-channel fully differential digital integrated neural recording and stimulation interface. IEEE Trans. Biomed. Circuits Syst. **4**(3), 149–161 (2010)

13. C.M. Lopez, A. Andrei, et al., An implantable 455-active-electrode 52-channel CMOS neural probe. IEEE J. Solid State Circuits **49**(1), 248–261 (2014)

14. M.S.J. Steyaert, W.M.C. Sansen, C. Zhongyuan, A micropower low-noise monolithic instrumentation amplifier for medical purposes. IEEE J. Solid State Circuits **22**(12), 1163–1168 (1987)

15. S. Song, M.J. Rooijakkers, et al., A low-voltage chopper-stabilized amplifier for fetal ECG monitoring with a 1.41 power efficiency factor. IEEE Trans. Biomed. Circuits Syst. **9**(2), 237–247 (2015)

16. T. Denison et al., A 2.2μW 100nV/√Hz, chopper-stabilized instrumentation amplifier for chronic measurement of neural field potentials. IEEE J. Solid State Circuits **42**(12), 2934–2945 (2007)

17. N. Verma et al., A micro-power EEG acquisition SoC with integrated feature extraction processor for a chronic seizure detection system. IEEE J. Solid State Circuits **45**(4), 804–816 (2010)

18. J. Xu, R.F. Yazicioglu, et al., A 160μW 8-channel active electrode system for EEG monitoring. IEEE Trans. Biomed. Circuits Syst. **5**(6), 555–567 (2011)

19. Y. M. Chi, G. Cauwenberghs, Micropower non-contact EEG electrode with active common-mode noise suppression and input capacitance cancellation, *Proc. IEEE EMBC*, (Sept. 2009), pp. 4218–4222

20. A. C. Metting-van Rijn et al., Low-cost active electrode improves the resolution in biopotential recordings. *Proc. IEEE EMBC*, (Oct. 1996), pp. 101–102

21. S. Mitra, J. Xu et al., A 700μW 8-channel EEG/contact-impedance acquisition system for dry-electrodes, *Digest of Symp. VLSI Circuits*, (June. 2012), pp. 68–69

22. M. Guermandi, R. Cardu et al., Active electrode IC combining EEG, electrical impedance tomography, continuous contact impedance measurement and power supply on a single wire, *Proc. ESSCIRC*, (Sept. 2011), pp. 335–338

23. J. F. Witte, J. H. Huijsing, K. A. A. Makinwa, A current-feedback instrumentation amplifier with 5μV offset for bidirectional high-side current-sensing, *Digest of ISSCC*, (Feb. 2008), pp. 74–75

24. B.J. van den Dool, J.H. Huijsing, Indirect current feedback instrumentation amplifier with a common-mode input range that includes the negative rail. IEEE J. Solid State Circuits **28**(7), 743–749 (1993)

25. S. Othmer, S.F. Othmer, D.A. Kaiser, J. Putman, Endogenous neuromodulation at Infralow frequencies. Semin. Pediatr. Neurol. **20**(4), 246–257 (2013)

26. ActiveTwo. [online] available: http://www.biosemi.com/activetwo_full_specs.htm

27. N. Van Helleputte, M. Konijnenburg et al., A multi-parameter signal-acquisition SoC for connected personal health applications, *Digest of ISSCC*, (Feb. 2014), pp. 314–315

28. R. Muller et al., A 0.013mm² 2.5μW, DC-coupled neural signal acquisition IC with 0.5V supply. IEEE J. Solid State Circuits **47**(1), 232–243 (2012)

Chapter 3
An Active Electrode Readout Circuit

3.1 IC Architecture Overview

The proposed EEG readout circuit (Fig. 3.1) consists of eight front-end AEs implemented with eight chopper IAs and one back-end voltage summing amplifier for CMFB.

The front-end AEs are responsible for transparent pre-amplification of EEG signals. To achieve this goal, several popular design techniques of bio-amplifiers are combined. Firstly, the AE utilizes a capacitive feedback IA architecture for rail-to-rail tolerance of electrode offset. Secondly, chopper modulation is performed at the amplifier's virtual ground to mitigate $1/f$ noise [1]. Thirdly, the AE includes an input impedance boosting loop for high input impedance [2]. Lastly, a ripple reduction loop (RRL) and a DC servo loop (DSL) compensate the intrinsic non-idealities of the chopper IA [3].

The back-end summing amplifier (Fig. 3.1) is responsible for CMRR improvement between multiple AEs. This amplifier performs common-mode (CM) signal extraction and feeds the input CM voltage of all eight AEs back to their non-inverting inputs (via V_{CMFB}). As a result, the CMFB scheme reduces the CM gain of these AEs for a high CMRR (see Sect. 3.3).

3.2 Active Electrode ASIC

The AEs, modeled as single-ended IAs, must achieve balanced performance, i.e., balancing input impedance, electrode offset tolerance, noise, CMRR, and power, to facilitate dry-electrode EEG recording. However, state-of-the-art IAs are not well

This chapter is derived from a journal publication of the authors: J. Xu, R.F. Yazicioglu, et al., "160μW 8-channel active electrode system for EEG monitoring," *IEEE Trans on Biomed Circuits and Systems*, vol. 5, no. 6, pp. 555–567, Dec. 2011.

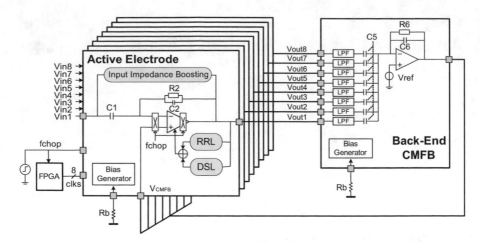

Fig. 3.1 Block diagram of the eight-electrode readout circuit

suited for this application. The capacitively-coupled IA [2] has limited input impedance formed by input switched-capacitor resistor. Moving the chopper to the amplifier's virtual ground solves this problem at the cost of degraded CMRR [1]. A current
feedback IA has good input impedance and CMRR, but its DC servo loop limits the
maximum electrode offset tolerance to 50 mV.

This section proposes a capacitively-coupled chopper IA similar to [1], with
inherent capability for rail-to-rail offset rejection and low integrated noise in a
100 Hz bandwidth. Furthermore, several additional circuit techniques are employed
to enhance the IA's input impedance, output dynamic range, and CMRR. Detailed
implementations of the core IA, including the offset trimming loops (RRL and DSL)
and the impedance boosting loop, are discussed in Sects. 3.2.1, 3.2.2, and 3.2.3,
respectively.

3.2.1   An AC-Coupled Inverting Chopper Amplifier

The IA's voltage gain is determined by the ratio of its feedback capacitors C_1/C_2
(Fig. 3.1). Variable gains (3, 10, 50, and 100) can be realized by switching between
different values of C_2. The pseudo-resistor R_2 and capacitor C_2 determine the AE's
high-pass cutoff frequency [4]. The coupling capacitor C_1 rejects any electrode
offset in a power efficient manner. Furthermore, the IA (Fig. 3.2) must have a
large output dynamic range to accommodate motion artifacts and interference.
Therefore, the core IA utilizes a folded cascode OTA, known for a good balance
between output voltage swing and power consumption.

Chopper modulation is used to achieve low noise. The input modulator is placed
before input transistors (M_1 and M_2), up-modulating the IA's input signals. The
output modulation is performed at the low impedance nodes before the dominant

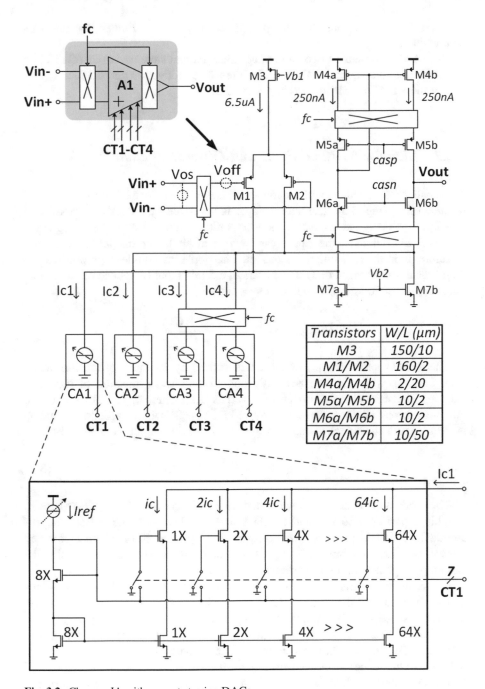

Fig. 3.2 Chopper IA with current steering DACs

pole (at V_{out}), such that the chopping frequency is not limited by the IA's bandwidth [5].

The IA also consists of two pairs of auxiliary current steering DACs (CA_1–CA_4) to compensate the chopper IA's non-idealities. The motivation and detailed operation of these DACs are discussed in Sect. 3.2.2.

3.2.2   Digitally Assisted Ripple and Offset Reduction

Two challenges associated with a chopper IA are how to reduce its output ripple and residual offset (Fig. 3.3). The output ripple is generated by the IA's up-modulated offset voltage (V_{off}) and resembles a low-pass filtered square wave. Compared to the μV level EEG signals, the ripple can have a much larger magnitude and will, therefore, limit the IA's output headroom. The ripple magnitude is proportional to the IA gain, as shown in (3.1). For instance, a 5 mV input offset can cause a large output ripple of 500 mV$_{pp}$ (when $C_1/C_2 = 100$).

$$\left| V_{ripple,pp} \right| = \left| V_{off} \right| \cdot \left(\frac{C_1 + C_2}{C_2} \right) \tag{3.1}$$

The residual offset $V_{off,out}$ of the IA is caused by its input offset current I_{os} (Fig. 3.3) [6], which in turn is mainly due to the charge injection of the input chopper. The residual output offset is derived as (3.2), where I_{os} is the offset current that flows through the pseudo-resistor R_2 and R_p, C_i is IA's input capacitance, and R_p is the IA's parasitic switched-capacitor resistor formed by input chopper and C_i.

$$V_{off,out} = I_{os}R_2 = \frac{V_{os}R_2}{R_p} = f_c C_i R_2 V_{os} \tag{3.2}$$

The residual offset can be compensated by a DC servo loop (DSL) [7], where an off-chip capacitor (>10 μF) and an OTA realize a low-pass cutoff frequency of around 0.5 Hz. In the proposed AE, however, the ripple and the offset are suppressed by two foreground calibration loops: a RRL and a DSL. It should be noted that the EEG input signal should not be present during the calibration.

The calibration starts with the RRL (Fig. 3.4): the ripple V_a and V_b are synchronously sampled, and the polarity (CMP$_1$) is determined by a comparator. A fully-integrated successive approximation algorithm (SAR) generates a pair of 7-bit binary outputs (CT_1 and CT_2) to control a pair of 7-bit current steering DACs (CA_1 and CA_2 in Fig. 3.2), respectively. The outputs of the SAR have inverse polarity, so that either a segment from the left DAC (CA_1) or from the right DAC (CA_2) is switched on after each comparison. Therefore, the DACs generate compensation currents (Ic_1 and Ic_2) to minimize the output ripple in seven clock cycles. The timing of the RRL's operation is illustrated in Fig. 3.5.

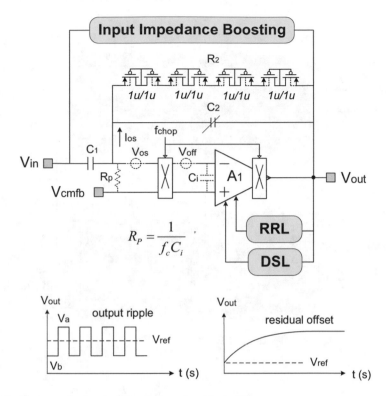

Fig. 3.3 Chopper-associated output ripple and residual offset

The DSL starts after the RRL and operates in a similar manner (Fig. 3.4): the output offset voltage (V_{out}) is sampled and compared to a reference voltage V_{ref}. The comparator output (CMP$_2$) is sent to the SAR, whose outputs control another pair of DACs (CA_3 and CA_4 in Fig. 3.4). Their outputs are chopper modulated to generate a modulated compensation current. The timing of the DSL's operation is also illustrated in Fig. 3.5.

Once both the RRL and the DSL calibration are finished, the inputs to the DACs are frozen, both calibration loops are shut-down, and normal operation starts. In addition, the calibration loops can be reset when necessary, in case there is any offset drift. The total power dissipation (<400 nW) of the RRL and the DSL is determined by the DAC's static current.

3.2.3 Input Impedance Boosting

Without an input impedance boosting loop, the input impedance of the inverting IA is dominated by C_1 (Fig. 3.6). This is illustrated in (3.3), where C_p is the input parasitic capacitance of the IA and R_s is the electrode-tissue impedance.

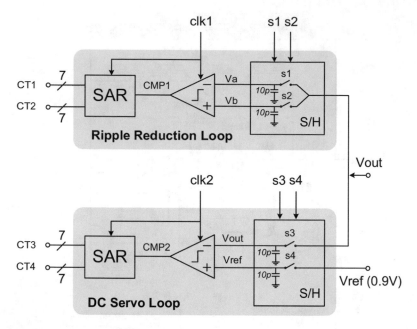

Fig. 3.4 Block diagrams of the RRL and the DSL

$$Z_{in} = \frac{1}{s(C_1 + C_p)} + R_s \approx \frac{1}{sC_1} \tag{3.3}$$

In case C_1 is large (300 pF) for low-noise operation (see Sect. 3.2.4), the input impedance is around 10 MΩ at 50 Hz. This may lead to a poor CMRR when an electrode impedance mismatch exists. In order to increase the input impedance, a positive feedback loop (Fig. 3.6) is implemented by feeding back an input bias current [2]. This loop consists of an inverting amplifier and a capacitor C_{fb}, which includes C_{fb_coarse} and C_{fb_fine}. C_{fb} converts the inverted output into an input bias current (I_{fb}), which is a portion of the total input current I_{in}. Therefore, the current (I_{el}) drawn from the recording electrode is reduced.

$$Z_{in} = \frac{V_{in}}{I_{el}} = \frac{1}{s(C_1 + C_p) - sC_{fb}\left[\frac{s^2 C_1 C_4 R_2 R_3}{(sC_2 R_2 + 1)(sC_3 R_3 + 1)} - 1\right]} \tag{3.4}$$

The input impedance Z_{in} of the AE, after utilizing the impedance boosting loop, can be expressed as in (3.4). Compared to (3.3), the equivalent input impedance has been increased by a factor of β, as shown in (3.5).

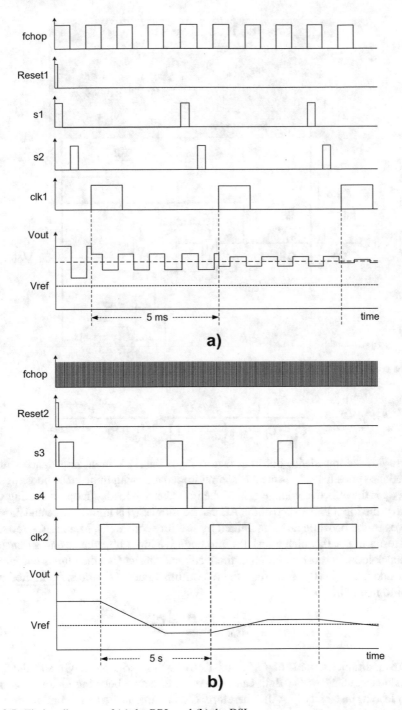

Fig. 3.5 Timing diagrams of (**a**) the RRL and (**b**) the DSL

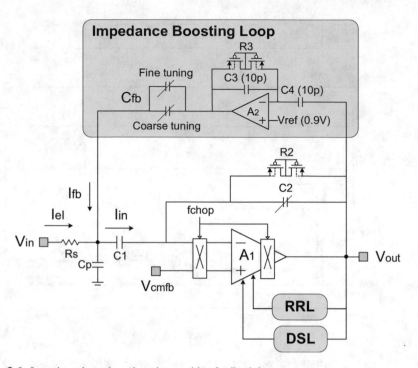

Fig. 3.6 Input impedance boosting via a positive feedback loop

$$\beta = \frac{Z'_{in}}{Z_{in}} = \frac{C_1 + C_p}{C_1 + C_p - C_{fb}\left(\frac{C_1 C_3}{C_2 C_4} - 1\right)} \approx \frac{C_1 + C_p}{C_1 + C_p - A_{V,IMP} C_{fb}} \qquad (3.5)$$

where β is the impedance boosting factor, Z_{in} is the AE's input impedance without impedance boosting, Z'_{in} is the AE's input impedance with impedance boosting, and $A_{V,AMP}$ is the effective voltage gain of the impedance boosting loop (excluding C_{fb}). Ideally, the input impedance of the AE can be infinite (β is infinite large and $I_{el} = 0$). However, the boosting factor is limited by stability constrains. Making C_{fb} too large will result in $\beta < 0$, which translates into negative input impedance and an unstable feedback loop, because a portion of the feedback current (I_{fb}) then flows out into the electrode (i.e., $I_{el} < 0$). Therefore, the maximum value of C_{fb} must be limited as in (3.6) to maintain $\beta > 0$:

$$C_{fb,max} = \frac{C_1 + C_p}{A_{V,IMP}} \qquad (3.6)$$

An additional remark of (3.5) and (3.6) is the variation of C_1 and the input parasitic capacitances C_p. Both can reduce the effective boosting factor β and may even lead to instability ($\beta < 0$). Therefore, C_{fb} is implemented as a combination of a coarse and a fine capacitor array to be able to trim the amount of positive feedback to compensate the effects of these process variation and ill-defined parasitic capacitance. At variable gain settings, the coarse array C_{fb_coarse} is switched in tandem with the value of C_2. The fine array C_{fb_fine} can then be adjusted to further compensate for the current that flows into C_1 and C_p, thus ensuring that β is high enough to guarantee stability. The selected C_{fb_fine} array (25 pF) can compensate for a 20% variation in C_1 (at the lowest gain of 3) and tolerate a large C_p of up to 15 pF.

3.2.4 Noise Analysis

The equivalent circuit for input noise derivation is shown in Fig. 3.7; the total input referred noise-power density of a front-end AE (FEAE) can be derived as

$$
\begin{aligned}
\overline{V^2_{in,FEAE}} &= \left(\overline{V^2_{in,OTA1}} + \overline{V^2_{in,cmfb}}\right) \cdot \left(1 + \frac{C_2}{C_1} + \frac{2\pi f_c C_i}{sC_1}\right)^2 \\
&+ \overline{V^2_{in,FEAE}} \cdot \left(\frac{1}{sC_1R_2}\right)^2 + \left(\overline{V^2_{in,OTA2}} + \overline{V^2_{in,ref}}\right) \cdot (2sR_sC_{fb})^2 \\
&+ \left(\overline{V^2_{in,RRL}} + \overline{V^2_{in,DSL}}\right) \cdot \left(\frac{g_{m,DAC}}{g_{m1}}\right)^2
\end{aligned}
\tag{3.7}
$$

where $V_{in,FEAE}$ is the total input referred noise of an AE; $V_{in,OTA1}$ and $V_{in,OTA2}$ are the input referred noise of the amplifier A_1 and A_2, respectively; $V_{in,cmfb}$ is the common-mode noise of the back-end CMFB amplifier; $V_{n,R2}$ is the noise contribution of the pseudo-resistor R_2; $V_{in,ref}$ is the noise of the reference voltage, which biases the inverting amplifier in the impedance boosting loop; $V_{in,RRL}$ and $V_{in,DSL}$ are the noise from the RRL and the DSL, respectively; $g_{m,DAC}$ is the transconductance of the current steering DACs; and g_{m1} is the input transconductance of the core amplifier A_1.

The noise of the impedance boosting loop ($V_{in,OTA2}$ and $V_{in,ref}$) is negligible as long as $1/sC_{fb} \gg R_s$. The noise generated from the pseudo-resistor R_2 is also very small as $sC_1R_2 \gg 1$. The noise of the RRL and the DSL is not dominant either because $g_{m1} \gg g_{m,DAC}$. The noise from the CMFB loop is common-mode noise for multiple AEs. Hence, the total input referred noise of a single AE can be approximated as

$$
\overline{V^2_{in,FEAE}} = \overline{V^2_{in,OTA1}} \cdot \left(1 + \frac{C_2}{C_1} + \frac{2\pi f_c C_i}{sC_1}\right)^2
\tag{3.8}
$$

This noise approximation is equal to the thermal noise of the core chopper IA, multiplied by a shaping factor (Fig. 3.8). This factor has its origins in the fact that the combination of the input chopper and the input capacitor C_i behaves like a parasitic switched-capacitor resistor, and so $V_{in,FEAE}$ exhibits a $1/f^2$ frequency characteristic. However, this approximation does not include the current noise contribution from the input chopper. The current noise can be converted into significant $1/f^2$ voltage noise as well when chopping is performed at a very high-impedance node, i.e., at the virtual ground of this inverting IA. A detailed discussion of this current noise can be found in Chap. 5.

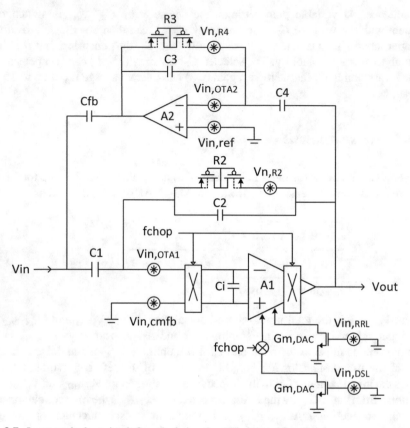

Fig. 3.7 Input equivalent circuit for calculating the AE's input referred noise

Several design guidelines should be considered to reduce the noise-shaping factor in (3.8). A large coupling capacitor C_1 should be used as long as the input impedance still meets the design requirements. The chopping frequency f_c should be selected very close to the $1/f$ corner of the core IA. In this design, a minimal f_c of 500 Hz is selected without significantly compromising the noise floor of the chopper IA. Moreover, the input parasitic capacitor C_i can be reduced by careful layout.

3.3   Back-End CMFB IC

Mismatch between AEs usually dominates their CMRR. This can be improved by a back-end CMFB circuit. Figure 3.9 shows the equivalent circuit of a simplified two-AE system and its CMFB circuit. Without the CMFB, the reference inputs of AEs are connected to ground. Therefore, the CM gain is determined by the AEs' gain mismatch (ΔA_V), leading to a low CMRR as shown in (3.9).

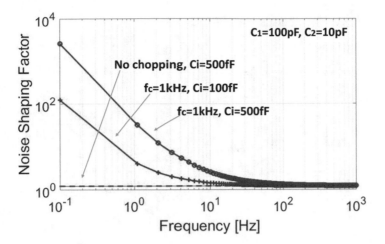

Fig. 3.8 Noise shaping factors on various conditions of chopping frequency (f_c) and differential input parasitic capacitance (C_i)

$$\text{CMRR} = 20\log\left(\frac{A_V}{\Delta A_V}\right) \tag{3.9}$$

With the CMFB, however, the reference inputs of the AEs are connected to the output of the CMFB, which is approximately equal to the input CM of all AEs. Thanks to the CMFB, it reduces the CM gain, the residual CM outputs V_{out1} and V_{out2}, as well as the differential output (V_{out}). The new CMRR', by using the CMFB, is derived in (3.10), where the CMRR is improved by a factor of $20\log(A_V)$. In (3.10), A_V is the close-loop voltage gain of the AE, and $A_{V,CMFB}$ is the CM gain of the capacitive summing amplifier.

$$\text{CMRR}' = \text{CMRR} \cdot 20\log\left(\frac{2A_V}{2 + \dfrac{1}{A_{V,CMFB}}} + 1\right) \tag{3.10}$$

$$\approx \text{CMRR} \cdot 20\log(A_V)$$

The back-end CMFB circuit (Fig. 3.10) consists of a capacitive summing amplifier with a gain of $A_{V,CMFB} = 8 \times (C_5/C_6)$ and eight unit-gain low-pass filters to reject high-frequency interference. The coupling capacitors (C_5) block the residual output offsets of the AEs, thus avoiding summing amplifier's saturation. Via a pseudo-resistor R_6, the summing amplifier provides a bias voltage to (V_{ref}) to all AEs. In an eight-electrode EEG readout circuit, any two AEs can be used to form a bipolar EEG acquisition channel. The summing amplifier only feeds the average CM voltage of all eight AEs back to their reference inputs while rejecting any differential-mode (DM) signals. Therefore, the back-end CMFB circuit does not disturb the differential EEG amplification.

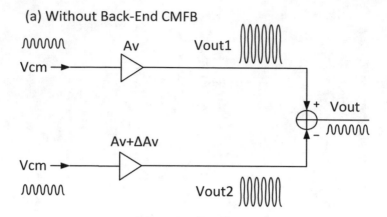

(a) Without Back-End CMFB

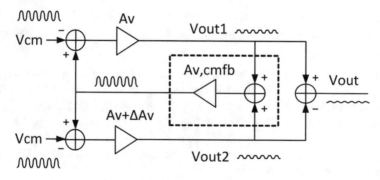

(b) With Back-End CMFB

Fig. 3.9 Equivalent circuits of the proposed AE system: (**a**) without CMFB and (**b**) with CMFB

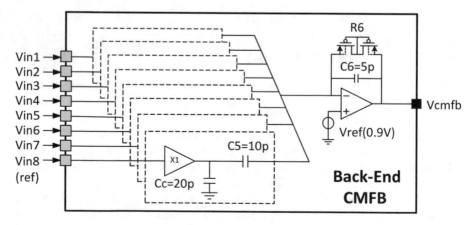

Fig. 3.10 Block diagram of the back-end CMFB circuit

The practical CMRR improvement is limited by the stability constrains of the CMFB loop. To balance the tradeoff between CMRR and the loop stability, the voltage gain of the AEs and the summing amplifier is set to 100 and 16, respectively.

In practice, electrode impedance mismatch is another mechanism further reducing the CMRR of a pair of AEs, especially when dry electrodes are used. In the case of CMFB, not being used (Fig. 3.11), the CMRR of an AE pair is derived in (3.11),

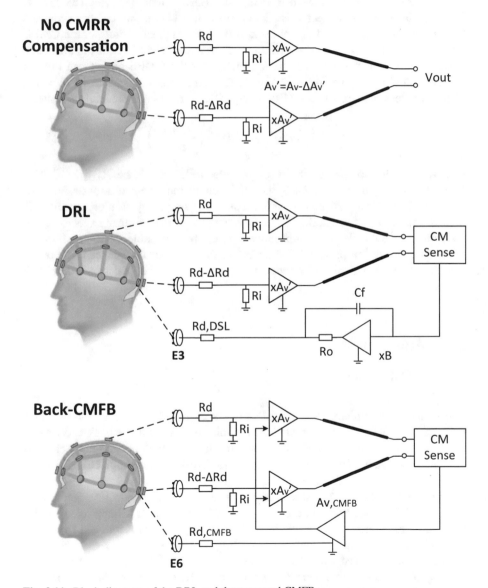

Fig. 3.11 Block diagrams of the DRL and the proposed CMFB

where ΔR_d is the electrode impedance mismatch and ΔA_V represents AEs' components mismatch.

$$\text{CMRR} = 20\log\left(\frac{A_V}{\Delta A_V + \frac{A_V \Delta R_d}{R_d + R_i}}\right) \qquad (3.11)$$

Compared to the well-known driven-right-leg (DRL) circuit [8], which feeds back the CM signal to the subject (Fig. 3.11) through a bias electrode, the proposed CMFB circuit feeds the CM signal back to the AEs' inputs. Therefore, the CMFB circuit only compensates for the AEs' components mismatch. Even if the mismatch is perfectly minimized, the maximum CMRR that the CMFB circuit can help to achieve will ultimately be limited by the electrode impedance mismatch, as shown in (3.12), whereas the maximum CMRR with a DRL circuit is theoretically infinite.

$$\text{CMRR}_{\text{CMFB,MAX}} = 20\log\left(\frac{R_d + R_i}{\Delta R_d}\right) \qquad (3.12)$$

However, due to the variability of the electrode impedance $R_{d,\text{DSL}}$ (10 kΩ–10 MΩ) and the stray capacitance, the DRL circuit must be carefully designed to ensure that it is always stable [9], which requires a large compensation capacitor C_f (a few tens of nF) for stability [10]. Dry electrodes may further exacerbate the instability since the electrode impedance is even more variable. In contrast, this is not an issue for the proposed CMFB circuit, as the electrode impedance is out of the feedback loop, and a third electrode (E_6) always biases the subject to the circuit ground.

3.4   Measurement

3.4.1   IC Measurement

The IC has been implemented in a 0.18 μm standard CMOS process and occupies 6.5 mm^2 (Fig. 3.12). Each fabricated die contains one AE and one back-end CMFB amplifier. An eight-electrode EEG readout circuit can be built up with eight chips as AEs and a separate chip operating as back-end CMFB. The eight-electrode readout circuit consumes 160 μW from 1.8 V.

Figure 3.13 shows the measured AE gain as it changes from 3 to 100. Figure 3.14 shows the AE's input referred noise with and without chopping. Chopping at 500 Hz leads to an integrated noise of 0.8μV$_{\text{rms}}$ (0.5–100 Hz), which is reduced by almost half compared to 1.5μV$_{\text{rms}}$ without chopping. Figure 3.15 shows that the input impedance is improved from 400 MΩ at 1 Hz to 2 GΩ at 1 Hz by utilizing the impedance boosting loop. Figure 3.16 shows the CMRR of a pair of AEs (with a gain of 100). An 82 dB CMRR has been measured at 50 Hz after enabling the back-end CMFB, which improves the initial CMRR by more than 30 dB. Figure 3.17 shows

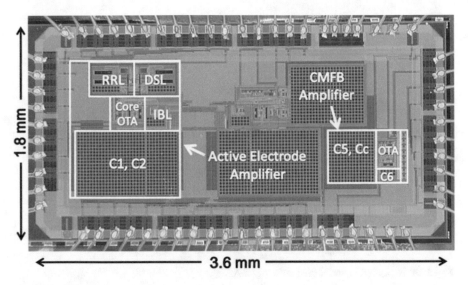

Fig. 3.12 Chip photograph

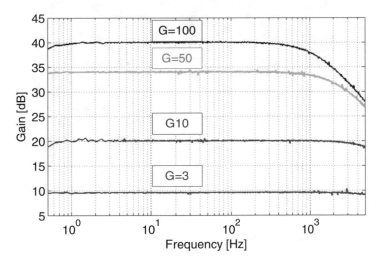

Fig. 3.13 Measured gain of an AE as a function of frequency for various gain factors (G)

the output waveforms before and after applying ripple and offset trimming. The residual output ripple is less than 2 mV compared to the initial 40 mV, and the output offset is reduced from 280 to 20 mV.

Table 3.1 summarizes the analog performance of the AE system and compares it with state-of-the-art AEs. The proposed AE system achieves the highest input impedance, comparable input referred noise, electrode offset rejection, and CMRR. The problem of achieving a high CMRR between single-ended AEs is

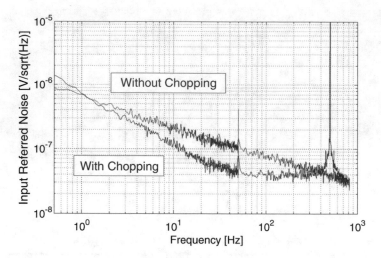

Fig. 3.14 Measured input referred noise of an AE

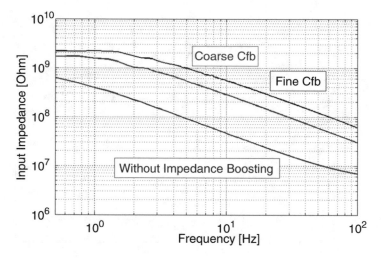

Fig. 3.15 Measured input impedance of an AE

essentially solved by using a back-end CMFB circuit. All these features make the proposed AE system suitable for dry-electrode EEG recording.

3.4.2   Cable Motion and Interference

This section demonstrates the benefits of an AE system by comparing its performance with a conventional EEG readout circuit. The latter is implemented with two

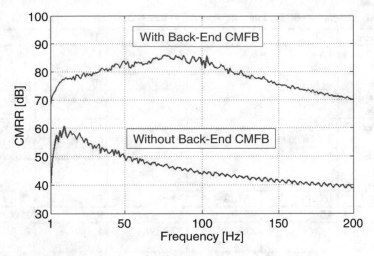

Fig. 3.16 Measured CMRR between an AE pair

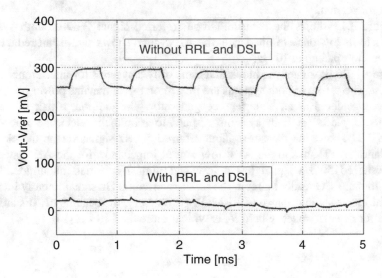

Fig. 3.17 Measured output ripple and residual offset of an AE

passive electrodes and a differential IA. The comparison includes their robustness to cable motion artifacts and interference. In Fig. 3.18, two resistors of 1 MΩ are placed at the IA's and the AEs' inputs to mimic the dry-electrode impedance. A low-noise ($3\mu V_{pp}$ in 0.1–10 Hz), high input impedance (2 GΩ), and high CMRR (>90 dB) IA [13] is selected as a conventional EEG IA. The cables connected between this IA and EEG electrodes are attached to a vibration device that vibrates constantly at 10 Hz. A similar measurement setup is used for the AE system proposed in this chapter, while

Table 3.1 Comparison between the proposed AE system and prior-art biopotential IAs

Parameters	[11]	[7]	[12]	[1]	This work
Supply voltage	1.8 V	3 V	1 V	1 V	1.8 V
AE gain	100	10	190–1000	100	3–100
Input impedance (DC)	>7.5 MΩ	>100 MΩ	–	>700 MΩ	2 GΩ
Noise per channel	0.95μV_{rms} (0.05–100 Hz)	0.6μV_{rms} (0.5–100 Hz)	2.5μV_{rms} (0.05–460 Hz)	1.3μV_{rms} (0.5–100 Hz)	0.8μV_{rms} (0.5–100 Hz)
Electrode offset tolerance	50 mV	50 mV	Rail-to-rail	Rail-to-rail	Rail-to-rail
CMRR	100 dB	120 dB	71 dB	60 dB	82 dB
NEF	4.6	9.2	3.3	9.5	12.3
Power per channel	2 μW	33.3 μW	337 nW	3.5 μW	20 μW (AEs only)

the IA is connected to the AEs' outputs through cables for differential to single-ended conversion.

Figure 3.19 shows the measured input referred signal PSD of both systems. Thanks to its low output impedance, the AE system shows significant reduction of cable motion pickup at 10 Hz.

Figure 3.20 illustrates the block diagrams of two systems for interference reduction test. The cables of both systems are placed on top of a mains power plug which can be considered as an "interference generator." In addition, a pair of variable resistors is placed at both systems' inputs to mimic the electrode impedance. Figure 3.21 shows the measured input-referred 50 Hz signal versus the electrode impedance R_S. Thanks to the AEs' low output impedance, for the AE system, the input-referred 50 Hz signal has a low and almost constant magnitude. In the conventional EEG readout circuit, the input-referred 50 Hz signal linearly increases with the electrode impedance (till 1 MΩ). Particularly, the benefits of AEs are more valued for dry electrodes, where R_S is typical more than 10 kΩ [14].

3.4.3   Biopotential EEG Measurement

Figure 3.22 shows the scalp EEG measurement setup. For simplicity, EEG signals are measured between two pairs of electrodes, which are both placed in positions of O_1 and C_z. A first pair is connected to two AEs ($G = 100$) via short wires. The outputs of these AEs are connected to a commercial g.tec EEG system (channel 1) [15]. For comparison, another pair of electrodes is placed very close to the first pair. The outputs of these electrodes are directly connected to the same EEG system (channel 2). In this way, the EEG measurement results can be compared simultaneously between the two types of systems.

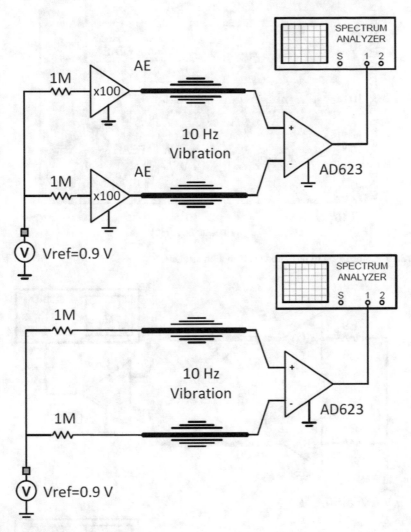

Fig. 3.18 Cable motion artifacts reduction test by introducing cable vibration

In the first measurement, both systems use wet electrodes. Figure 3.23 shows the spectrogram of the measured EEG signal. Alpha waves around 10 Hz are clearly visible when eyes are closed. In Fig. 3.24, the spectrum correlation coefficient (ρ) between the (wet electrode) AE readout and the (wet electrode) conventional EEG readout is higher than 0.99.

In the second measurement, wet-electrode AEs are replaced with dry-electrode AEs [16], while the commercial EEG readout channel still uses wet electrodes. Figure 3.25 shows the EEG spectrogram. For both systems, alpha waves around 10 Hz are still clearly visible when eyes are closed. In addition, Fig. 3.26 shows that the spectrum correlation coefficient (ρ), between the dry-electrode AE readout and

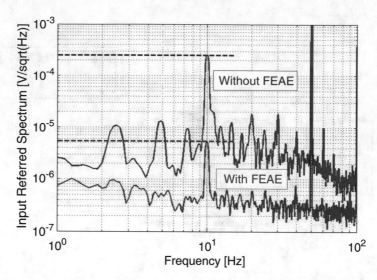

Fig. 3.19 Comparison of measured cable motion artifacts

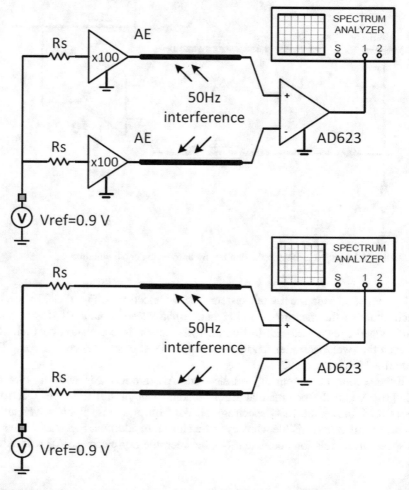

Fig. 3.20 Interference reduction test by introducing a 50 Hz interference source

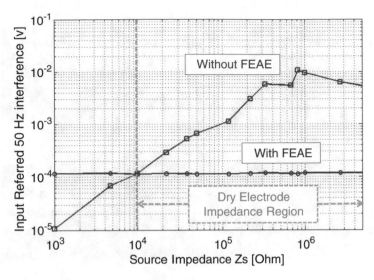

Fig. 3.21 Comparison of measured interference at 50 Hz

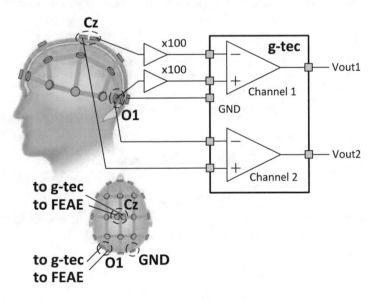

Fig. 3.22 Scalp EEG measurement setup

the wet-electrode commercial EEG readout, is higher than 0.93. This number is high in comparison to other works on dry-electrode sensing, such as [17]: $\rho > 0.9$, [18]: $\rho = 0.8$–0.96, and [19]: $\rho = 0.83$.

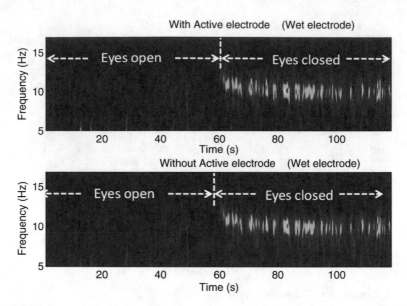

Fig. 3.23 EEG spectrogram with AE (upper trace, wet electrodes) and without AE (bottom trace, wet electrodes)

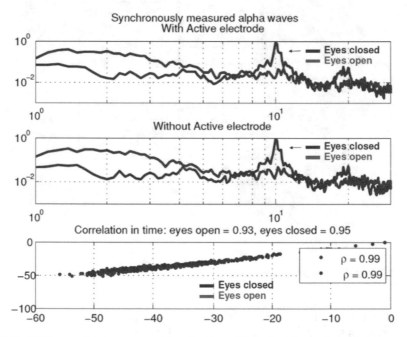

Fig. 3.24 EEG normalized spectrum with and without AE (both with wet electrodes)

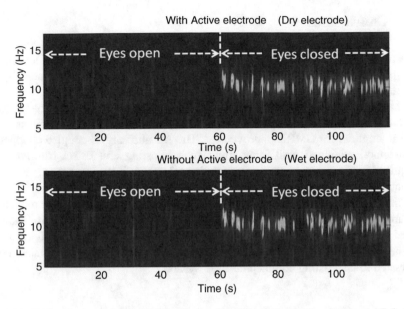

Fig. 3.25 EEG spectrogram with AE (upper trace, dry-electrode headset) and without AE (bottom trace, wet electrode)

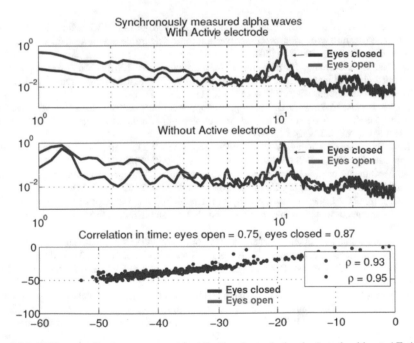

Fig. 3.26 EEG normalized spectrum with AE (dry-electrode headset) and without AE (wet electrode)

3.5   Conclusions

A wearable EEG system requires both user-friendly dry electrodes and low-power high-performance readout circuits. However, state-of-the-art IAs, implemented with differential amplifiers, are not well suited for such applications.

This chapter presented a low-power AE-based readout circuit for dry-electrode EEG measurement. The readout circuit includes eight IAs as AEs and one back-end CMFB amplifier for CMRR improvement. The AE utilizes AC-coupled chopper IA architecture, equipped with input impedance boosting and offset trimming, for optimized performance between noise, offset tolerance, input impedance, and large output swing.

The AE-based readout circuit shows significant benefits in terms of the robustness to cable motion and interference than a traditional EEG readout circuit. The proposed AE system can detect alpha waves when either wet or dry electrodes are used on the scalp. Moreover, this AE system also shows a highly correlated result compared to an existing commercial EEG system.

References

1. N. Verma, A. Shoeb, A.J. Bohorquez, et al., A micro-power EEG acquisition SoC with integrated feature extraction processor for a chronic seizure detection system. IEEE J. Solid State Circuits **45**(4), 804–816 (2010)
2. Q. Fan, F. Sebastiano, H. Huijsing, K.A.A. Makinwa, A 1.8µW 60nV/√Hz capacitively-coupled chopper instrumentation amplifier in 65nm CMOS for wireless sensor nodes. IEEE J. Solid-State Circuits **46**(7), 1534–1543 (2011)
3. R. Wu, K.A.A. Makinwa, J.H. Huijsing, A chopper current-feedback instrumentation amplifier with a 1mHz 1/f noise corner and an AC-coupled ripple reduction loop. IEEE J. Solid State Circuits **44**(12), 3232–3243 (2009)
4. R.R. Harrison, C. Charles, A low-power low-noise CMOS amplifier for neural recording applications. IEEE J. Solid State Circuits **38**(6), 958–965 (2003)
5. M. Sanduleanu et al., A low noise, low residual offset, chopped amplifier for mixed level applications. Proc. IEEE Int. Conf. Electron Circuits Syst. **2**, 333–336 (1998)
6. C.C. Enz, G.C. Temes, Circuit techniques for reducing the effects of op-amp imperfections: Autozeroing, correlated double sampling, and chopper stabilization. Proc. IEEE **84**, 1584–1614 (1996)
7. R.F. Yazicioglu, P. Merken, R. Puers, et al., A 60µW 60nV/√Hz readout front-end for portable biopotential acquisition systems. IEEE J. Solid State Circuits **42**(5), 1100–1110 (2007)
8. B.B. Winter, J.G. Webster, Driven-Right-Leg circuit design. IEEE Trans. Biomed. Eng. **30**(1), 62–66 (1983)
9. T. Degen, H. Jackel, Enhancing interference rejection of preamplified electrodes by automated gain adaption. IEEE Trans. Biomed. Eng. **51**(11), 2031–2039 (2004)
10. T. Jochum, T. Denison, P. Wolf, Integrated circuit amplifiers for multi-electrode intracortical recording. J. Neural Eng. **6**(1), 012001 (2009)
11. T. Denison, K. Consoer, A. Kelly et al., A 2.2µW 94nV/√Hz, chopper-stabilized instrumentation amplifier for EEG detection in chronic implants, *Digest of ISSCC*, (Feb. 2007), pp. 162–594

12. X. Zou, W. Liew, L. Yao, L. Yong, A 1V 450nW fully integrated programmable biomedical sensor interface chip, IEEE J. Solid-State Circuits, 44(4) (Apr. 2009), pp. 1067–1077
13. AD623. [online] available: http://www.analog.com/static/imported-files/data_sheets/AD623.pdf
14. Y.M. Chi, T.-P. Jung, G. Cauwenberghs, Dry-contact and noncontact biopotential electrodes: Methodological review. IEEE Rev. Biomed. Eng. **3**, 106–119 (2010)
15. g.USBamp. [online] available: http://www.gtec.at/Products/Hardware-and-Accessories/g.USBamp-Specs-Features
16. L. Brown, J. van de Molengraft, R. F. Yazicioglu, T. Torfs, J. Penders, C. Van Hoof, A low-power, wireless, 8-channel EEG monitoring headset, *IEEE EMBC*, (Aug. 2010), pp. 4197–4200
17. R. Matthews, P. J. Turner, N. J. McDonald, K. Ermolaev, T. Mc Manus, R. A. Shelby, M. Steindorf, Real time workload classification from an ambulatory wireless EEG system using hybrid EEG electrodes, *IEEE EMBC*, (Aug. 2008), pp. 5871–5875
18. J.R. Estepp, J.C. Christensen, J.W. Monnin, I.M. Davis, G.F. Wilson, Validation of a dry electrode system for EEG. Human Factors and Ergonomics Society Annual Meeting **53**(18), 1171–1175 (2009)
19. G. Gargiulo, P. Bifulco, R. A. Calvo, M. Cesarelli, C. Jin, A. van Schaik, A mobile EEG system with dry electrodes, *IEEE Biomedical Circuits and Systems Conference*, (Nov. 2008), pp. 273–276

Chapter 4
An Eight-Channel Active Electrode System

4.1 IC Architecture Overview

Figure 4.1 shows the block diagram of the proposed eight-channel EEG/ETI acquisition system. An EEG measurement is obtained as the differential output of two AEs and so the system consists of nine AEs. A bias electrode provides a DC bias (at ½ V_{dd}) for all the AEs. Each AE connects to the BE via six wires: 32 kHz clock (CLK), power supply (VDD and VSS), digital control bits (PWM), analog output (ANA) of EEG and ETI, and common-mode feedforward (CMFF). The AE utilizes a non-inverting chopper IA for pre-amplification and a built-in square-wave current source for ETI measurement [1].

The BE is responsible for analog signal processing and digitization. In Fig. 4.2, the BE signal chain starts with two chopper IAs, each consisting of a transconductance (TC) stage and a transimpedance (TI) stage. The two TI stages in an ETI channel are used to demodulate and separate the EEG signal from the ETI signals, and each ETI channel consists of an inphase (I) channel and a quadrature (Q) channel. Programmable gain amplifiers (PGA) provide variable gain, and low-pass filters (LPF) enable anti-aliasing. Both EEG and ETI channels are simultaneously sampled at 500 Hz by respective sample-and-hold (S/H) stages. The result is a total of 24 channels, including eight-channel EEG, eight-channel ETI-I, and eight-channel ETI-Q. These outputs are multiplexed and digitized by two 12-bit SAR ADCs operating at 1 kS/s.

This chapter is derived from a journal publication of the authors: J. Xu, S. Mitra, et al., "A 700 µW 8-Channel Active Electrode EEG/Contact-impedance Acquisition System," *IEEE J. Solid-State Circuit,* vol.49, no.9 pp. 2005–2016, Sept. 2014.

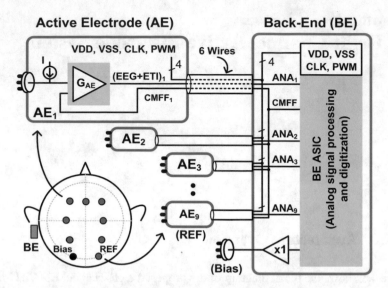

Fig. 4.1 Block diagram of an eight-channel AE-based EEG/ETI system

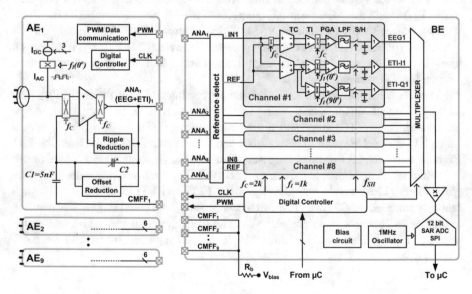

Fig. 4.2 Architecture of the eight-channel EEG/ETI acquisition system (TC and TI represent transconductance and transimpedance IAs, respectively)

4.1.1 EEG and ETI Measurement

The proposed eight-channel system can simultaneously measure EEG and ETI signals (Fig. 4.3). A DC current (I_{DC}) is up-modulated to the ETI measurement frequency ($f_I = 1$ kHz) and injected into the subject through each EEG recording

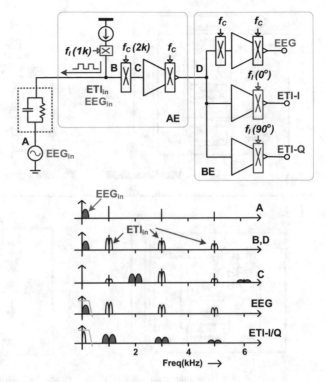

Fig. 4.3 Block diagram and spectrum to illustrate the principle of EEG/ETI measurement

electrode. This current is then converted into a square-wave voltage (ETI_{in}) over the electrode impedance. Thus, the EEG and ETI signals are both present at the input of an AE, but are located at different frequencies. At the output of an EEG channel, the amplified EEG signal is still at baseband, and the residual ETI signal at f_I can easily be removed by a LPF. The ETI-I/Q channels up-modulates the EEG signal to f_I and filters it out, while they demodulate inphase f_I ($0°$) and quadrature f_I ($90°$) components of the ETI signal back to DC, respectively. Typically, the ETI-I is much larger than the ETI-Q at the ETI measurement frequency (at a few kHz). For instance, 51 kΩ//47 nF, the standard (wet) electrode impedance, has ETI-I = 51 kΩ and ETI-Q = 3.3 kΩ at 1 kHz, respectively.

4.1.2 A CMFF Technique for CMRR Enhancement

Gain mismatch between AEs typically limits their intrinsic CMRR to less than 60 dB. A DRL circuit can improve CMRR at the cost of potential instability [2]. The back-end CMFB proposed in the previous chapter can solve this issue by feeding the CM signal back to the AEs' reference inputs, instead of the subject. However, the CMFB requires a summing amplifier for CM extraction and uses large capacitors for stability, leading to increased chip area and power. Alternatively, a

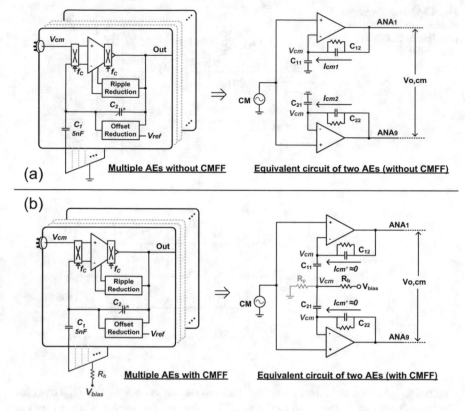

Fig. 4.4 (**a**) Conventional AEs without CMFF and equivalent circuit diagram of two AEs (**b**) AEs with CMFF for CMRR enhancement and equivalent circuit of two AEs

digitally-assisted DRL technique [3] improves its stability, where digital notch filtering ensures that high loop gain is only available at the major interference frequency (at 50/60 Hz). However, this improvement is at the expense of complex digital signal processing and more power.

This section introduces a CMFF technique for CMRR improvement of multiple AEs [4]. Conventionally, the non-inverting AEs are referred to the system ground (Fig. 4.4a). Consequently, the CMRR of a pair of AEs is limited by their voltage gain error, as shown in (4.1), where G_{AE} is the voltage gain of an AE, and ΔG_{AE} is the gain difference of two AEs. Although the matching of AEs can be improved by implementing both C_1 and C_2 with a common-centroid layout with dummies; the maximum CMRR will still be typically less than 60 dB. This issue can be solved by a CMFF technique, where the AEs' reference node (now dubbed the CMFF node). This node was previously connected to the circuit ground, and now it is capacitively connected to a DC reference voltage (V_{bias}) through a very large bias resistor R_b (Fig. 4.4b). Thus, the CMFF node becomes an averaging node for all the input signals, and, as a result, its voltage is equal to the CM input. Furthermore, no CM current will flow through capacitors C_{11} and C_{21}. This effectively reduces the CM gain and thus increases the CMRR of an AE pair, as shown in (4.2), where CMRR is the initial CMRR of two AEs without using the CMFF, C_1 is the input capacitor (C_{11} or C_{21}) and R_b is the bias resistor.

$$CMRR = 20\log\left(\frac{G_{AE}}{\Delta G_{AE}}\right) \qquad (4.1)$$

$$CMRR' = CMRR + 20\log(1 + sC_1R_b) \qquad (4.2)$$

However, increasing R_b will make the DC voltage at the CMFF node sensitive to leakage due to the voltage divider formed by R_b and the parasitic resistance R_p (Fig. 4.4b). Furthermore, it will significantly reduce the amplitude of the DC voltage at CMFF node, thus limiting the maximum CM swing that the CMFF can handle. As a reasonable requirement, the actual DC voltage at the CMFF node should be larger than the AE's maximum input CM voltage ($V_{in,pk}$), as given by

$$V_{DC,CMFF} = V_{bias} \cdot \frac{R_p}{R_p + R_b} > |V_{in,pk}| \qquad (4.3)$$

Equation (4.3) sets an upper limitation of R_b. On the other hand, decreasing R_b will limit the maximum CMRR that the CMFF can achieve, as shown in (4.2). A small R_b increases the high-pass frequency at the CMFF node and thus prevents any CM voltage extraction. For instance, when $R_b = 0$, the CMFF will be biased to ground, and the CMFF will stop working. In this design, $V_{bias} = 1.8$ V and $R_b = 100$ MΩ are selected to accommodate a CM input range of 50 mV while maintain a good CMRR (>80 dB).

One remaining limitation of the CMFF is its robustness to the "lead-off" condition. Disconnecting any electrode will cause the failure of the CMFF loop, because a floating input of any AE pollutes the CM averaging [4]. This problem can be solved by connecting the positive input of each AE to a well-defined bias voltage (at $V_{dd}/2$) via a large resistor. Thus, the AE in the lead-off condition will be biased, while CMFF is then performed among the other AEs.

4.1.3 PWM Communication

In an AE-based system, each AE receives the configuration register bits from the BE to define its operation modes, such as the AE's gain and bandwidth, and the amplitude and frequency of the ETI current source. Data communication between the BE and the AEs could be done via a 3-wire SPI interface. However, this would lead to an increased number of wires in a multichannel system, increasing the system's complexity and weight. In this design, the BE-to-AE interface utilizes a single-wire self-clocked PWM data transmission [5], which combines the clock and data signals.

4.2   Active Electrode ASIC

4.2.1   Instrumentation Amplifier

As the analog front-end of the EEG system, an AE should provide high input impedance and low noise, as well as the capability to reject large electrode offsets. To meet these requirements, the AE (Fig. 4.5) consists of a non-inverting chopper IA equipped with a DC servo loop (DSL) and a ripple reduction loop (RRL). Their working principle and design tradeoffs are discussed in the following sections in detail.

Input Impedance. A non-inverting IA has much higher input impedance than that of an inverting IA [6] because the input impedance of the former architecture is determined by input parasitic capacitance, instead of the large coupling capacitor (C_1). In this design (Fig. 4.5), the IA has an input impedance of 2 GΩ at 20 Hz, while

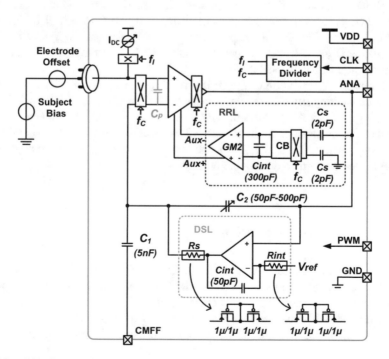

Fig. 4.5 Block diagram of the IA used as an AE

the output resistance ($R_{out,IDC}$) of the ETI current source is 3.2 GΩ at 20 Hz. This gives the AE an equivalent input impedance of more than 1.2 GΩ at 20 Hz.

$$Z_{in} = \left(\frac{1}{2f_c C_p} + \frac{1}{sC_1} \right) // R_{out,IS} \tag{4.4}$$

Electrode Offset Rejection. To reject a large DC electrode offset (DEO), which may otherwise saturate the AE, the built-in IA implements a continuous-time DSL (Fig. 4.5). The DSL stabilizes the output DC voltage to a reference voltage (V_{ref}) via an active RC integrator, which feeds the IA's output offset back to its inverting input via a large pseudo resistor (R_s). The maximum electrode offset rejection of each AE is ±250 mV with respect to the subject bias. This is ultimately determined by the input biasing range of the core amplifier.

Ripple Reduction. The intrinsic offset of the core amplifier will be up-modulated by the output chopper. This generates a square wave, namely, ripple, thus reducing the IA's output headroom. A RRL (Fig. 4.5) [7] first converts the output ripple into a DC current via a capacitor (C_s) and a chopped current buffer (CB). This DC current is then integrated on a capacitor (C_{int}), and the resulting voltage is converted into a DC current via a transconductance stage (GM2). This DC current compensates the IA's intrinsic offset and reduces the output ripple approximately by a factor of 10.

4.2.2 Noise Analysis

In this work, the input-referred noise specification of an EEG channel (including two AEs and one BE) is 75 nV/sqrt(Hz). Neglecting the noise contribution from the BE, the input-referred noise of a single AE should be about 53 nV/sqrt(Hz), which includes three major contributors (Fig. 4.6): the core amplifier, the RRL, and the DSL.

$$\overline{V_{in,AE}^2} = \overline{V_{in,IA}^2} + \overline{V_{in,RRL}^2} + \overline{V_{in,DSL}^2} \tag{4.5}$$

The noise contribution of the core IA can be expressed by

$$\overline{V_{in,IA}^2} = \overline{V_{in,OP1}^2} \cdot \left[1 + \frac{C_2}{C_1} + \left(\frac{f_c}{f_{in}} + 1 \right) \cdot \frac{C_p}{C_1} \right]^2 + \overline{I_{in,OP1}^2} \cdot \left(\frac{1}{2\pi f_{in} C_2} \right)^2 \tag{4.6}$$

where $V_{ni,OP1}$ and $I_{ni,OP1}$ are the input-referred voltage noise and current noise of the core amplifier, respectively, C_1 and C_2 are the feedback capacitors which define the AE's gain, f_c is the chopping frequency, f_{in} is the frequency of the input signal, and C_p is the input parasitic capacitance.

The first term in (4.6) refers to the $1/f^2$ voltage noise of the non-inverting chopper IA because of its parasitic switched-capacitor resistance. Although chopper

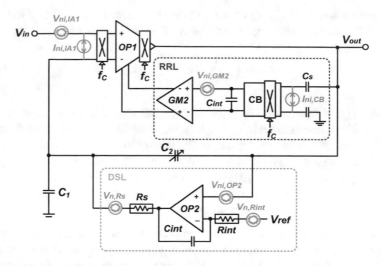

Fig. 4.6 Equivalent noise model of the AE

modulation mitigates $1/f$ noise of the core amplifier (OP1), this parasitic resistance reduces the input impedance of the amplifier and thus increases its current noise. The current noise is converted into $1/f^2$ voltage noise in the presence of external capacitive feedback. The second term in (4.6) refers to another $1/f^2$ voltage noise source associated with the current noise $I_{ni,OP1}$, which is due to the charge injection and clock feedthrough of the input chopper [8]. The current noise PSD linearly increases with the chopping frequency [9] and induces significant $1/f^2$ voltage noise, especially when chopping is performed at a high-impedance node. This problem will be discussed in more detail in Chap. 5.

The proposed IA utilizes several methods to minimize the total noise in (4.6). Firstly, to reduce the thermal noise of $V_{ni,OP1}$, the core amplifier employs a two-stage amplifier (Fig. 4.7), whose input pair consists of NMOS and PMOS differential pairs connected in series [10]. This method doubles the input transconductance of the core amplifier without consuming extra bias current while still achieving a good input CM range from 0.7 to 1.2 V. Secondly, in order to reduce $I_{ni,OP1}$, the input chopper modulator utilizes a low chopping frequency ($f_c = 2$ kHz) and small-size transistors [8]. Thirdly, the use of a non-inverting topology means its input impedance is not dominated by the feedback capacitors. Thus, very large capacitors C_1 (5 nF) and C_2 (50–500 pF) are used to reduce the impedance of the chopping node, as shown in (4.6). This topology reduces the $1/f^2$ voltage noise without compromising IA's input impedance. This is a clear advantage of a non-inverting chopper AE, as the inverting IA in Chap. 3 suffers from the tradeoff between noise and input impedance [6, 11].

The RRL's noise contribution is negligible since it is located between the two choppers of the core amplifier (OP1). The $1/f$ noise from GM2 and current buffer (CB) are effectively chopped out, while their thermal noise is suppressed by the input stage of OP1. The DSL does not induce significant noise in the EEG bandwidth either, as its low-pass cutoff frequency is well below 0.5 Hz.

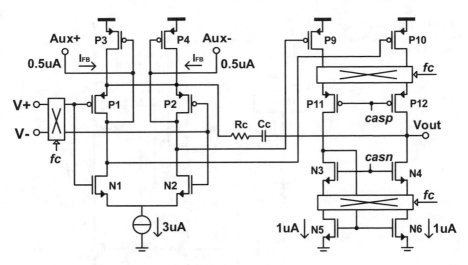

Fig. 4.7 Schematic of the core amplifier of an AE

4.2.3   Current Source for ETI Measurement

The current source employs a self-biased triple-cascode architecture to boost its output impedance (>3.2 GΩ at 20 Hz) (Fig. 4.8). The magnitude of the reference current (I_{dc}) is configurable from 10 nA to 2 μA to cover a wide range of ETI. This reference current is mirrored either to a NMOS or PMOS triple-cascode stage, enabling current injection or current sink. An output chopper then generates a square-wave current at the ETI measurement frequency ($f_I = 1$ kHz).

4.3   Back-End Analog Signal Processing ASIC

4.3.1   Instrumentation Amplifiers

In the BE, one EEG channel and two ETI channels use the same IA architecture but with different chopping schemes to separate EEG and ETI from each other. The IA of the EEG channel (Fig. 4.2) utilizes both input and output choppers to reduce its 1/f noise, while two ETI channels share the same TC stage of the EEG channel for low power and use only output choppers in their TI stages to demodulate the ETI signal.

The IA is built with a current-feedback instrumentation amplifier (CFIA) (Fig. 4.9) [12]. The TC stage employs input voltage followers for high input impedance. The differential input voltage across the input resistor R_i creates a signal dependent current. This current is mirrored (via P_8 and P_6) to the TI stage and converted back to a voltage through the output resistor R_o. The voltage gain of the IA is given by

Fig. 4.8 Triple-cascode, self-biased, square-wave current source

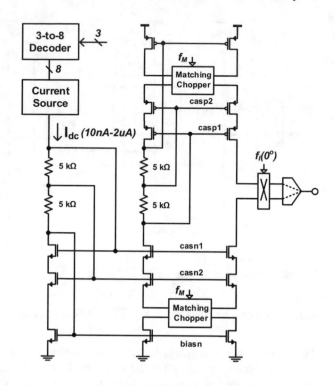

Fig. 4.9 Schematic of TC and TI stages of IA in the BE

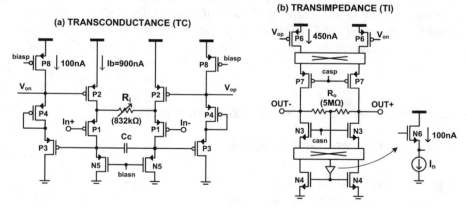

$$G_{\mathrm{IA}} = \frac{R_0}{R_i}\left(\frac{W_6/L_6}{W_2/L_2}\right) \tag{4.7}$$

The level shifters (P_3 and P_4) help to keep the voltages V_{on} and V_{op} at reasonable values for a wide range of input CM voltages. In this architecture, the maximum input swing of the IA is determined by $R_i I_b$.

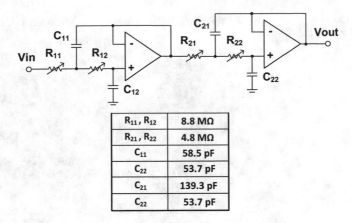

R_{11}, R_{12}	8.8 MΩ
R_{21}, R_{22}	4.8 MΩ
C_{11}	58.5 pF
C_{22}	53.7 pF
C_{21}	139.3 pF
C_{22}	53.7 pF

Fig. 4.10 Schematic of the fourth-order Sallen-Key low-pass filter (LPF)

4.3.2 Low-Pass Filter and ADC

The anti-aliasing LPFs separate the EEG signal and the ETI signal (at 1 kHz), and reject chopping spikes. The LPF is a fourth-order unity-gain Bessel filter realized with a Sallen-Key architecture (Fig. 4.10) [13]. This LPF provides a tunable band-width (100–300 Hz), as well as sufficient attenuation (>60 dB at 1 kHz) of residual ETI signal. In addition, the use of a Bessel filter ensures that the EEG and ETI channels all have a constant group delay of about 1 ms. The sample-and-hold (S/H) circuits thus sample all channels at more or less the same time. Both constant group delay and synchronized sampling improve the accuracy of temporal correlation across channels. This is important for brain-computer-interface (BCI) applications, where the ETI output can be used for impedance-related motion artifact reduction [14].

Two time-multiplexed 12-bit SAR ADCs, conceptually similar to the ones used in [15], digitize all 24 analog outputs (EEG, ETI-I, and ETI-Q from 8 channels) at 1 kS/s.

4.4 Measurement

The eight-channel AE system, including the AEs and the BE, is implemented in a 0.18 μm standard CMOS process. Figure 4.11 shows the photographs of both chips and the packaged AE placed on an 11 mm diameter electrode. The eight-channel system consumes less than 700 μW from 1.8 V. Table 4.1 shows the system power breakdown.

Figure 4.12 shows the measured PWM signal sent from the BE and the demodulated serial data received by the AE, which demonstrates a proper recovery of the register bits.

Fig. 4.11 Chip photographs: (left) the AE, (right) the BE

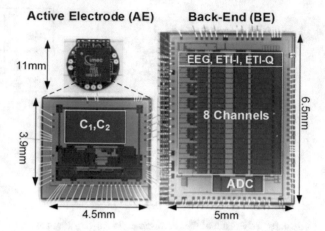

Table 4.1 Power breakdown of the eight-channel AE system

Active electrode (AE)	$11.1\,\mu A \times 9 = 100\,\mu A$
Core amplifier	$5\,\mu A$
DSL	$1.5\,\mu A$
RRL	$0.7\,\mu A$
Bias	$1.9\,\mu A$
Current source	$2\,\mu A$
Back-end (BE) readout	$265\,\mu A$
Bias	$10\,\mu A$
8-channel EEG	$56\,\mu A$
16-channel ETI	$83.2\,\mu A$
ADC + SPI	$115.2\,\mu A$
Total power of the system	$365\,\mu A @ 1.8\,V = 657\,\mu W$

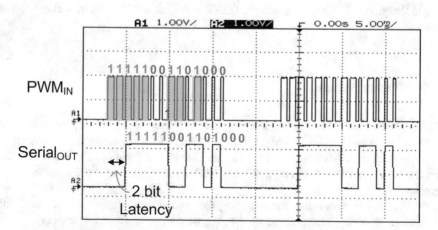

Fig. 4.12 Measured PWM input data (PWM_{IN}) and demodulated output ($Serial_{OUT}$)

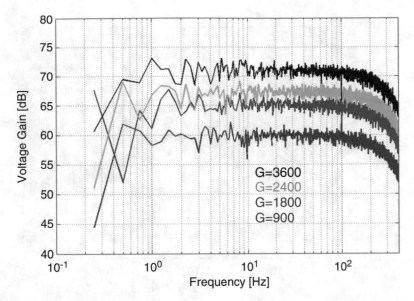

Fig. 4.13 Measured EEG channel gain as a function of frequency for various gain factors G (900, 1800, 2400, and 3600)

Fig. 4.14 Measured input impedance of an AE as a function of frequency

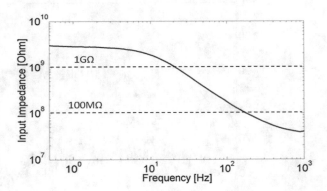

Figure 4.13 shows the measured voltage gain of one EEG channel at various PGA gain settings (3, 9, 12, and 18). The measured bandwidth of 200 Hz is determined by the LPF.

Figure 4.14 shows the measured input impedance of an AE, which is 1.2 GΩ at 20 Hz and above 300 MΩ at 50 Hz.

Figure 4.15 shows the input-referred noise of one EEG channel (including two AEs and one BE) versus frequency. When chopping is disabled, $1/f$ noise is clearly visible; when chopping is enabled at 2 kHz, $1/f^2$ noise dominates till 20 Hz. Above 20 Hz, the input noise density is constant at 75 nV/sqrt(Hz). The integrated noise is $1.75\mu V_{rms}$ from 0.5 to 100 Hz.

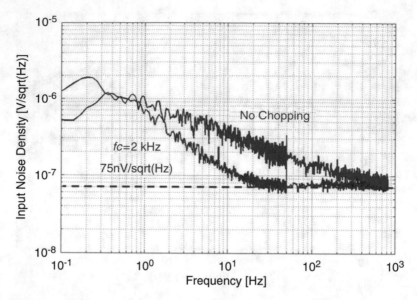

Fig. 4.15 Measured input-referred noise density per channel (two AEs and one BE)

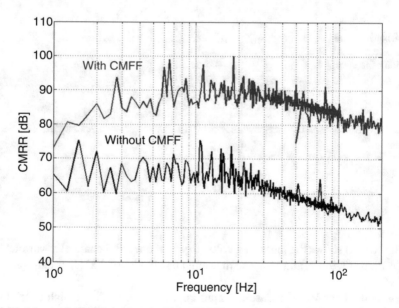

Fig. 4.16 Measured CMRR (with and without CMFF)

Figure 4.16 shows the measured CMRR of one EEG channel (including two AEs and one BE). In this measurement, the input CM signal is made of a 100 mV_{pp} chirp from 1 to 200 Hz, and it was applied to both AEs directly without any electrode impedance. At 50 Hz, enabling the CMFF improves the CMRR by 25 dB (from 60 to

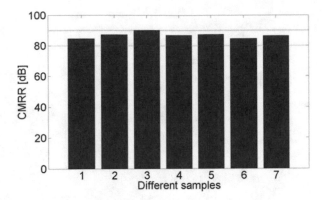

Fig. 4.17 Measured CMRR at 50 Hz from different samples

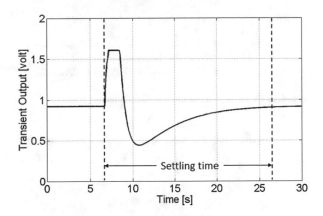

Fig. 4.18 Measured settling time of the IA

85 dB). Figure 4.17 shows that the CMRR between different pairs of AEs is always above 84 dB.

Figure 4.18 shows the AE's output voltage as a function of time. When a large transient electrode offset of 200 mV is applied to the input of AE, its output first saturates and then recovers to the reference voltage in about 20 s. This settling time is determined by the DSL and can be reduced by a fast setting buffer.

Figures 4.19 and 4.20 show the measured ETI resistance and capacitance versus their actual values. In these measurements, either a test resistor (110 Ω to 280 kΩ) or a test capacitor (100 pF to 1 µF) was connected to the input of one AE, while the other AE was connected to the subject bias voltage of $V_{dd}/2$ directly. The gain of the AE and BE was set to 101 and 9, so that both the EEG and ETI channels have the same gain ($\approx$900). In this default setting, the maximum differential ETI signal that the system can correctly measure is approximately 60 kΩ or 2 nF (at $f_I = 1$ kHz). This is limited by the output swing of the PGA in the BE (0.35–1.45 V), as well as the magnitude of the injected current (10 nA). By lowering the gains of the AE and BE to 11 and 9, respectively, ETIs of up to 550 kΩ can be measured at the expense of less gain ($\approx$100) in the EEG channel. In principle, even larger ETIs can be measured by lowering the magnitude of the injected current.

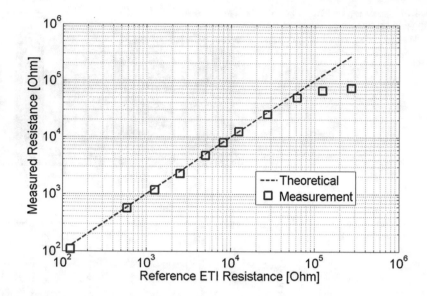

Fig. 4.19 Measured differential ETI resistance

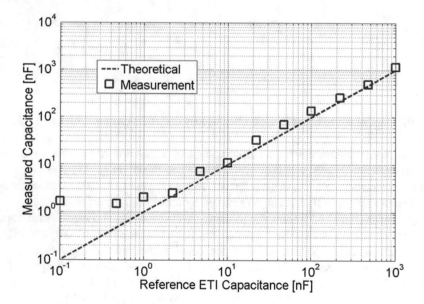

Fig. 4.20 Measured differential ETI capacitance

Table 4.2 compares the analog performance of the proposed eight-channel AE system with state-of-the-art AE systems. The proposed system is able to measure EEG and ETI signals simultaneously while still achieving very competitive performance on input impedance, noise, electrode offset tolerance, and CMRR. When it

Table 4.2 Comparison table with the state-of-the-art AE system

Parameters	[6] + [17]	[18]	[19]	[20]	This work
Technology	0.18 μm	N/A	0.35 μm	N/A	0.18 μm
Supply	1.8 V	5 V	3 V	3.3 V	1.8 V
AE gain	3–100	100	N/A	11	11–101
Input impedance	0.6GΩ@10 Hz	1TΩ@DC	N/A	N/A	1.2GΩ@20 Hz
Noise per channel	1.2μV$_{rms}$ (0.5–100 Hz)	7.49μV$_{rms}$ (1–1 kHz)	0.9μV$_{rms}$ (0.5–100 Hz)	2.4μV$_{rms}$ (0.5–100 Hz)	1.75μV$_{rms}$ (0.5–100 Hz)
Electrode off-set rejection	Rail-to-rail	±250 mV	N/A	Rail-to-rail	±250 mV
CMRR	82 dB	78 dB	105 dB	90 dB	84 dB
ETI measurement	No	No	Yes	No	No
Power per channel (including ADC)	20 μW (excl. [17])	7.5 mW	1 mW	600 μW	82 μW (EEG + ETI)

was published in [16], the proposed AE system also achieves the lowest power dissipation per channel.

4.5   A Four-Channel Wireless EEG Headset

The low-power highly integrated AEs and BE are well suited for a battery-powered wearable EEG device. A four-channel wireless EEG headset prototype using these chips has been implemented (Fig. 4.21). This headset consists of four recording AEs, one reference AE, and one bias electrode. Two mechanical bridges cover all the electrode positions and have compartments for sensor node electronics and battery. The recording AEs are positioned at predefined positions C_3, C_4, C_z, and P_z according to a 10–20 electrode EEG system. The reference AE and bias electrode are positioned behind the ears on the mastoid bone. All these electronics are connected via flat cables and embedded in the headset.

The digital outputs of BE ASIC are streamed out to a low-power microcontroller [21] via a SPI protocol and stored in local memory. The data is then transmitted wirelessly to a PC through a low-power radio [22]. The data transmission of microcontroller and radio occupies more than 80% of the system power, while the AEs and the BE only consume 5%.

Figure 4.22 shows the spectrogram of four-channel EEG signals measured from a subject whose eyes were alternately opened and closed. There is no alpha wave during the eyes-open period for all channels, except some artifacts due to blinking. During the eyes-closed period, the alpha waves at 10 Hz are clearly visible on all four channels.

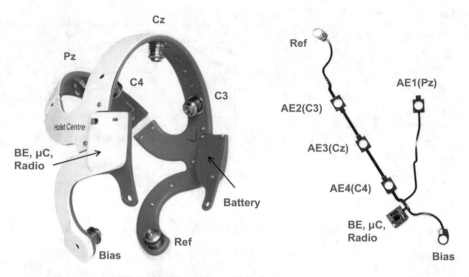

Fig. 4.21 Wireless EEG headset and its internal electronics

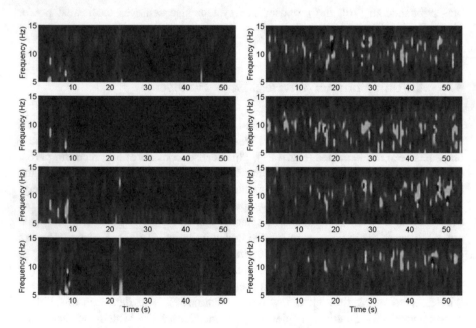

Fig. 4.22 4-channel real-time EEG recording using a dry-electrode wireless headset, during eyes open (left) and eyes closed (right). Output channels from top to bottom: C_3, C_4, C_z, P_z

4.6 Conclusions

This chapter presents a complete eight-channel active electrode (AE) system for simultaneous EEG/ETI measurement with dry electrodes. The whole system consists of nine AEs for high-performance pre-amplification and a low-power highly configurable BE for analog signal processing and digitization. Several techniques were implemented to improve the system performance. At the AE level, electrode offset is rejected by a low-power DSL around a non-inverting chopper IA, whose input stage utilizes improved transconductance for greater noise efficiency. At the system level, a CMFF technique improves the CMRR of an AE pair by 25 dB, a single-wire PWM modulation reduces the number of wires between the AEs and the BE, and a continuous-time ETI measurement can sense electrode impedances up to 550 kΩ. To demonstrate its functionality, the AE system was used to realize a battery-powered wireless EEG headset.

References

1. S. Kim, R.F. Yazicioglu, T. Torfs, B. Dilpreet, P. Julien, C. Van Hoof, A 2.4μA continuous-time electrode-skin impedance measurement circuit for motion artifact monitoring in ECG acquisition systems, *Symp. VLSI Circuits Digest*, (June 2010), pp. 219–220
2. B.B. Winter, J.G. Webster, Driven-Right-Leg circuit design. IEEE Trans. Biomed. Eng. **30**(1), 62–66 (1983)
3. M.A. Haberman, E.M. Spinelli, A multichannel EEG acquisition scheme based on single ended amplifiers and digital DRL. IEEE Trans. Biomed. Circuits Syst. **6**(6), 614–618 (2012)
4. A.C. Metting-van Rijn, A. Peper, C.A. Grimbergen, High-quality recording of bioelectric events. Part 2. Low-noise, low-power multichannel amplifier design. Med. Biol. Eng. Comput. **29**(4), 433–440 (1991)
5. J. Lime, S. Silva, A. Cordeiro, M. Verleysen, A CMOS/SOI single-input PWM discriminator for low-voltage body-implanted applications. VLSI Design **15**, 469–476 (2002)
6. J. Xu, R.F. Yazicioglu, P. Harpe, K.A.A. Makinwa, C. Van Hoof, A 160μW 8-channel active electrode system for EEG monitoring, *Digest of ISSCC*, (Feb. 2011), pp. 300–302
7. R. Wu, K.A.A. Makinwa, J.H. Huijsing, A chopper current-feedback instrumentation amplifier with a 1mHz 1/f noise corner and an AC-coupled ripple reduction loop. IEEE J. Solid State Circuits **44**(12), 3232–3243 (2009)
8. J. Xu, Q. Fan, J.H. Huijsing, C. Van Hoof, R.F. Yazicioglu, K.A.A. Makinwa, Measurement and analysis of current noise in chopper amplifiers. IEEE J. Solid-State Circuits **48**(7), 1575–1584 (2013)
9. D. Drung, J.-H. Storm, Ultralow-noise chopper amplifier with low input charge injection. IEEE Trans. Instrum. Meas. **60**(7), 2347–2352 (2011)
10. X. Zou et al., A 1V 22μW 32-channel implantable EEG recording IC, *Digest of ISSCC*, (Feb. 2010), pp. 126–127
11. N. Verma, A. Shoeb, A.J. Bohorquez, et al., A micro-power EEG acquisition SoC with integrated feature extraction processor for a chronic seizure detection system. IEEE J. Solid State Circuits **45**(4), 804–816 (2010)
12. R.F. Yazicioglu, P. Merken, R. Puers, et al., A 60μW 60nV/√Hz readout front-end for portable biopotential acquisition systems. IEEE J. Solid State Circuits **42**(5), 1100–1110 (2007)

13. T. Kugelstadt, Active filter design techniques, in *Op Amps for Everyone: Design Reference*, (Newnes, Boston, 2001), pp. 271–281
14. N. Van Helleputte, S. Kim, H. Kim, J.P. Kim, C. Van Hoof, R.F. Yazicioglu, A 160μW biopotential acquisition IC with fully integrated IA and motion artifact suppression. IEEE Trans. Biomed Circuits Syst. **6**(6), 552–561 (2012)
15. R.F. Yazicioglu, P. Merken, R. Puers, C. Van Hoof, A 200μW eight-channel EEG acquisition ASIC for ambulatory EEG systems. IEEE J. Solid State Circuits **43**(12), 3025–3038 (2008)
16. S. Mitra, J. Xu et al., A700μW 8-Channel EEG/Contact-impedance Acquisition System for Dry-electrodes, *Symp. VLSI Circuits Digest*, (June 2012), pp. 68–69
17. g.USBamp [Online] available: http://www.gtec.at/Products/Hardware-and-Accessories/g. USBamp-Specs-Features
18. T. Degen, H. Jackel, A pseudo differential amplifier for bioelectric events with DC-offset compensation using two-wired amplifying electrodes. IEEE Trans. Biomed. Eng. **53**(2), 300–310 (2006)
19. M. Guermandi, R. Cardu et al., Active electrode IC combining EEG, electrical impedance tomography, continuous contact impedance measurement and power supply on a single wire, *Proc of ESSCIRC*, (Sept. 2011), pp. 335–338
20. Y. Chi, G. Cauwenberghs, Micropower non-contact EEG electrode with active common-mode noise suppression and input capacitance cancellation, *IEEE EMBS*, (Sept. 2009), pp. 4218–4222
21. MSP430F161, Texas Instruments, [online] available: http://www.ti.com/product/msp430f1611
22. nRF24L01+, Nordic Semi [online] available: http://www.nordicsemi.com/kor/Products/2. 4GHz-RF/nRF24L01P

Chapter 5
Current Noise of Chopper Amplifiers

5.1 Chopping and Current Noise

Chopping is a continuous-time technique in which polarity-reversing switches, known as choppers, are used to modulate amplifier offset and $1/f$ noise to a certain chopping frequency, thus enabling the realization of precision amplifiers with low-voltage noise and low offset [1]. As a result, chopper amplifiers are often used in applications where precision signal conditioning is required, e.g., in smart sensors, sensor interfaces [2], medical instruments [3], and precision voltage references [4].

In CMOS, the chopper switches are usually implemented as MOSFETs. Although it is well known that the transient spikes caused by charge injection and clock feedthrough of these periodically switched devices will give rise to a net input current [5, 6], not much is known about the associated current noise. In [7], the current noise of a chopper amplifier was attributed to the shot noise associated with this input current. In [8], measurements of the current noise of a chopper amplifier are described. Although the cause of this noise was not explained, it was observed that the measured noise density was proportional to the square root of the chopping frequency and was in the order of several tens of fA/sqrt(Hz). This is roughly a hundred times higher than the current noise of conventional CMOS- or JFET-input amplifiers [9, 10]. Some commercially available chopper amplifiers exhibit even higher current noise ($\geq$100 fA/sqrt(Hz)) [11–13]. When used with high-impedance sensors such as dry electrodes, photodiodes, and piezoelectric sensors, this current noise will be converted to voltage noise, which will then add to and may even dominate the IA's total input-referred voltage noise [14].

This chapter is derived from a journal publication of the authors: J. Xu, Q. Fan, et al., "Measurement and Analysis of Current Noise in Chopper Amplifiers" *IEEE J. Solid-State Circuit*, vol.48, no.7, pp. 1575–1584, July. 2013.

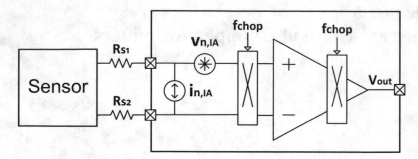

Fig. 5.1 Equivalent circuit model which describes the excess voltage noise due to the source impedance and input current noise of a chopper amplifier

5.2   Current Noise Analysis

Figure 5.1 shows the equivalent input circuit of a chopper IA connected to a differential voltage source. The IA's noise is modeled by an input-referred voltage noise source ($v_{n,IA}$) and an input-referred current noise source ($i_{n,IA}$), while the IA itself is considered to be ideal and noiseless. $R_{s1,2}$ models the source resistances. The total input-referred voltage noise can then be written as

$$\overline{V_{in,\,amp}^2} = \overline{V_{in,\,IA}^2} + \overline{I_{in,\,IA}^2} \cdot (R_{s1} + R_{s2})^2 + 4kT(R_{s1} + R_{s2}) \tag{5.1}$$

The input bias current of a CMOS amplifier is usually quite low (in the order of a few pA [9]) and is dominated by the gate leakage current of the input transistors and the leakage current of the ESD protection circuit. The associated current noise ($i_{n,IA}$) is mainly due to shot noise, and so it is also quite low (<1 fA/sqrt(Hz)) [9, 10]. Hence, the current noise of a CMOS amplifier is usually a negligible contributor to its total input-referred noise.

However, CMOS chopper amplifiers exhibit substantially higher levels of current noise [11–13]. This excess noise must therefore be related to the periodic switching of the MOSFET switches of the input chopper. The rest of this section presents an analysis of the major noise sources associated with this activity.

5.2.1   Charge Injection and Clock Feedthrough

Charge injection and clock feedthrough are well-known error sources associated with MOSFET switches. In a chopper, one pair of switches will be "on," while the other is "off." As shown in Fig. 5.2, when a pair of NMOS switches is turned off, their channel charge and some of the charge in their overlap capacitance (C_{ol}) will be injected into the circuitry connected to their drain and source terminals (modeled by the capacitors C_s and C_i) [15].

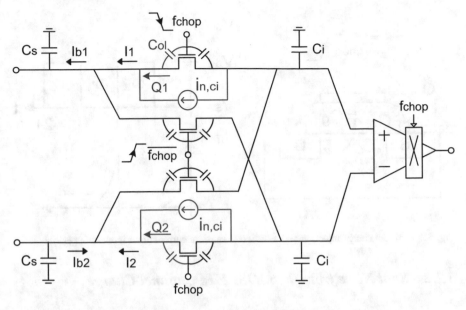

Fig. 5.2 Charge injection and clock feedthrough of the input chopper switches

The total charge ($Q_{1,2}$) that is injected into the source (or drain) circuit is given by (5.2), where W and L are the width and length of the chopper switches, C_{ox} is the gate oxide capacitance, C_{ol} is the overlap capacitance between gate and source (drain), V_{od} is the overdrive voltage, and V_{clk} is the clock swing.

$$Q_{1,2} = WLC_{ox}V_{od} + C_{ol}V_{clk} \qquad (5.2)$$

This periodic charge injection and clock feedthrough at the chopping frequency causes transient current spikes (Fig. 5.3), whose average value ($I_{1,2}$) is given by

$$I_{1,2} = \frac{Q_{1,2}}{T_{chop}/2} = 2f_{chop}(WLC_{ox}V_{od} + C_{ol}V_{clk}) \qquad (5.3)$$

Ideally, the transient current spikes caused by a pair of chopper switches turning off should be compensated by the charge required to turn on the other pair, leading to a net zero input current. However, mismatch between the switches and slight differences in their turn-on and turn-off times results in a net input current with a typical magnitude of several tens of pA [16, 17]. This is much larger than the gate leakage currents of the MOSFETs or the leakage currents of the ESD diodes. The right-hand side of (5.3) may thus be regarded as an upper bound, especially since the exact amount of input current will also depend on the relative magnitudes of the capacitances C_i and C_s connected to the chopper. From (5.3), the input current should be proportional to the chopping frequency, which is in good agreement with the measurements reported in [8].

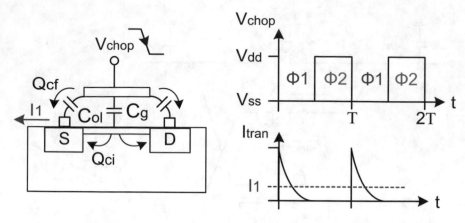

Fig. 5.3 Periodic charge injection and associated transient current of an input chopper switch

5.2.2   Shot Noise from the MOSFETs Channel Charge

Shot noise is associated with the nonuniform flow of charge carriers in semiconductors. This noise has a white noise spectrum, whose PSD is proportional to the average current [18, 19]. Since the current spikes associated with the charge injection of the MOSFETs in the periodically switched MOSFETs give rise to a net input current, the hypothesis is that this current will also be accompanied by shot noise [7]. The PSD of this current noise should then be linearly dependent on the average current $I_{1,2}$ through the chopper switches, as shown in (5.4), where $q = 1.6e^{-19}$ C is the electron charge. The average noise density of this impulse noise can then be expressed as

$$\overline{i_{n,\,ci}^2} \; \propto 2qI_{1,2} \quad \overline{i_{n,\,ci}^2} \; \propto 4qf_{\text{chop}}WLC_{ox}V^{ox}\text{od} \qquad (5.4)$$

This indicates that the current noise PSD associated with charge injection should also be linearly dependent on the chopping frequency.

5.2.3   KT/C Noise from the Clock Driver

The clock driver circuit is another possible source of noise. As shown in Fig. 5.4, it can be modeled as a resistance (R_g) in series with the gate-source capacitance (C_{gs}). Since this resistor (and any other series resistance in the gate charging circuit) will generate thermal noise, the channel charge will fluctuate, and so a certain noise charge will be injected into the surrounding circuitry every time the MOSFET is turned off. The root mean square (rms) value of this noise charge ($Q_{n,rms}$) can be expressed as

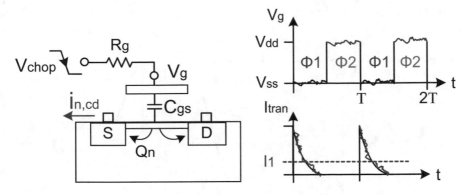

Fig. 5.4 Periodical noise charge injection and associated noise current of an input chopper switch

$$Q_{n,\text{rms}} = \sqrt{kTC_{\text{gs}}} = \sqrt{kTWLC_{\text{ox}}} \tag{5.5}$$

As before, this periodically injected noise charge will give rise to an average RMS noise current of

$$\overline{i_{n,\text{rms}}} = 2f_{\text{chop}}Q_{n,\text{rms}} = 2f_{\text{chop}}\sqrt{kTWLC_{\text{ox}}} \tag{5.6}$$

Assuming that this impulse noise is approximately white and distributed over the fundamental interval between 0 and $f_{\text{chop}}/2$, then its PSD is given by

$$\overline{i_{n,\text{cd}}^2} = \frac{\overline{i_{n,\text{rms}}^2}}{\Delta f} = 8f_{\text{chop}}kTWLC_{\text{ox}}^{\;ox} \tag{5.7}$$

This PSD is also a linear function of the chopping frequency.

5.2.4 Parasitic Switched-Capacitor Resistance

Due to the action of the input chopper, the amplifier's input parasitic capacitances (C_p) will be charged and discharged by the input voltage and give rise to a net DC current [5, 6]. As shown in Fig. 5.5, this effect can be modeled by a switched-capacitor resistor (R_{sc}) at the amplifier's input [20, 21]. This resistance generates current noise in the same manner as a physical resistor [22].

$$\overline{i_{n,\text{sc}}^2} = \frac{4kT}{R_{\text{sc}}} \qquad R_{\text{sc}} = \frac{1}{2f_{\text{chop}}C_p} \tag{5.8}$$

The resulting current noise PSD is again proportional to the chopping frequency (f_{chop}) and to the input parasitic capacitance (C_p) of the amplifier. Since the switched-capacitor resistor is usually quite large (tens or even hundreds of MΩ), the magnitude of the current noise PSD is usually negligible.

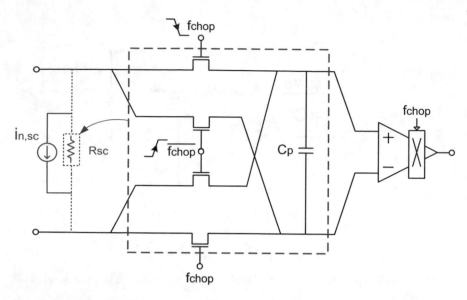

Fig. 5.5 Input parasitic SC resistance of a chopper amplifier

Table 5.1 Parameters of a typical MOSFET switch in 0.5 μm CMOS technology and calculated current noise-induced contribution to voltage noise density of a CMOS chopper consisting of eight MOSFETs ($T = 25$ °C or 298 K)

Parameters	Explanations	Typical value	Unit
W	Width	30	μm
L	Length	0.5	μm
C_{ox}	Gate oxide capacitance	2.5	fF/μm^2
V_{gs}	Gate-source voltage	1.9	V
V_{th}	Threshold voltage	0.7 (NMOS)	V
		0.9 (PMOS)	
f_{chop}	Chopping frequency	4	kHz
C_p	Input parasitic capacitance	125	fF
$i_{n,ci}$	Current noise density (charge injection)	30.4	fA/sqrt(Hz)
$i_{n,cd}$	Current noise density (kT/C noise)	1.1	fA/sqrt(Hz)
$i_{n,sc}$	Current noise density (parasitic SC resistance)	4.1	fA/sqrt(Hz)

5.2.5   *Summary*

The total chopper noise PSD ($i_{n,IA}$) is obtained by summing the contributions of all the abovementioned current noise sources. Table 5.1 shows the parameters of the MOSFET switches (in an ON Semiconductor 0.5 μm CMOS process) used in the input chopper of a CMOS chopper IA [3]. Also shown is the calculated contribution of each noise source, assuming that eight of these transistors (four NMOS and four PMOS) are used to realize the four complementary switches of its input chopper. The results show that the total current noise is dominated by the contribution of charge injection, and so, from (5.4), the chopper noise PSD should be linearly proportional to the chopping frequency.

5.3 Current Noise Measurement

5.3.1 A Conventional Chopper Modulated Amplifier

Figure 5.6 shows the schematic used to measure the current noise of a chopper IA. Its input chopper consists of four complementary CMOS switches, whose characteristics are shown in Table 5.1. The IA is configured with a voltage gain of 800 and a bandwidth of 200 Hz. Since it was intended for biomedical applications, an internal DC servo loop ensures that the amplifier has a high-pass characteristic with a corner frequency of approximately 0.5 Hz.

A low-noise input bias voltage, V_b, is generated from a 3.3 V battery. Two large resistors ($R_s = 10$ MΩ) ensure that the chopper IA's current noise is the dominant contributor to its total input-referred noise; i.e., that equations in (5.9) are satisfied [23].

$$\overline{i_{n,IA}^2}R_s^2 > \overline{v_{n,IA}^2} \quad \overline{i_{n,IA}^2}R_s^2 > 4kTR_s \tag{5.9}$$

The input current noise PSD can then be determined from (5.10), where $v_{n,out}$ is the measured output noise voltage, G is the IA's voltage gain, and the IA's voltage noise $v_{n,IA}$ and source resistance R_s are known. Note that the thermal noise of the choppers' on-resistance is included in the measured $v_{n,IA}$.

$$\overline{i_{n,IA}^2} = \left(\frac{\overline{v_{n,out}^2}}{G^2} - \overline{v_{n,IA}^2} - 8kTR_s \right) / 2R_s^2 \tag{5.10}$$

Figures 5.7 and 5.8 show the measured voltage gain (G) and the input-referred voltage noise density ($v_{n,IA}$) of the chopper IA, respectively. These results were used to determine the input-referred current noise density and confirm the proper operation of the IA at the various chopping frequencies.

The measured input-referred current noise PSD at various chopping frequencies is shown in Figs. 5.9 and 5.10. As predicted by (5.4), the PSD of this chopper noise is linearly proportional to f_{chop}. Figure 5.11 shows that the measured input current noise density is independent of the value of source resistance, as expected.

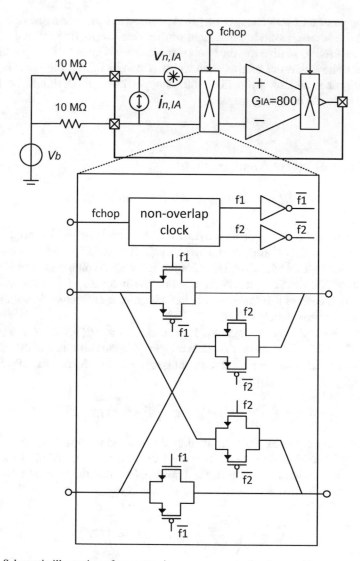

Fig. 5.6 Schematic illustration of current noise measurement of a chopper IA

Figure 5.12 shows the measured SC input impedance of the chopper IA at different chopping frequencies. The smallest input impedance is about 250 MΩ, which corresponds to the highest chopping frequency of 16 kHz. Hence, the maximum current noise density associated with this SC input impedance is only 8 fA/sqrt(Hz), which is negligible compared to the measured total current noise density of 158 fA/sqrt(Hz) at this chopping frequency.

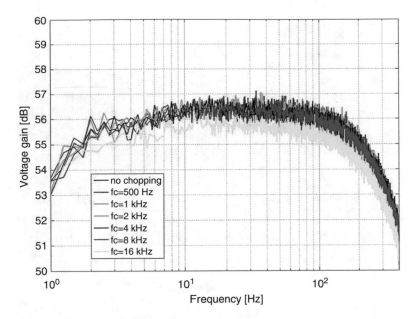

Fig. 5.7 Measured voltage gain (G) of the chopper IA

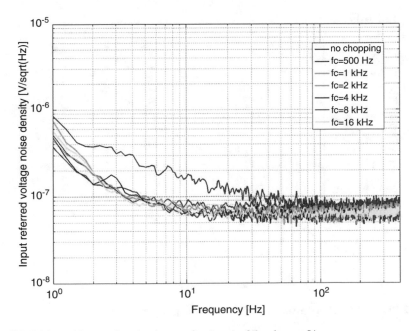

Fig. 5.8 Measured input-referred voltage noise ($v_{n,IA}$) of the chopper IA

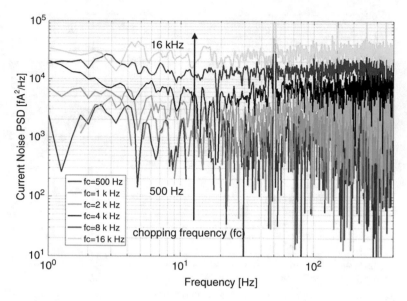

Fig. 5.9 The IA's input current noise PSD

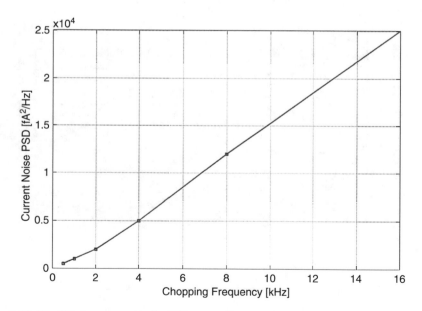

Fig. 5.10 The IA's input current noise PSD versus chopping frequencies

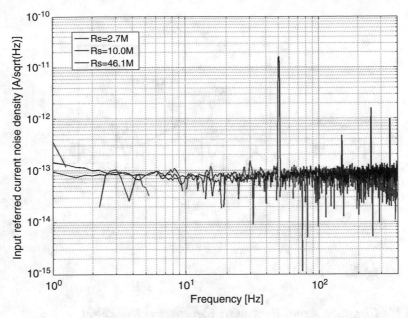

Fig. 5.11 The IA's input current noise density with different source resistors ($f_{\text{chop}} = 4$ kHz)

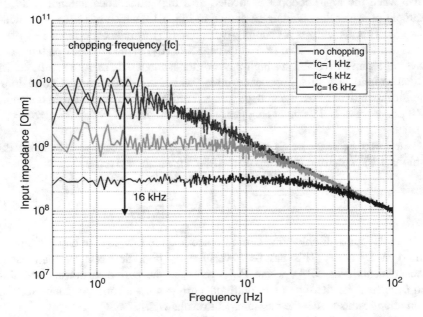

Fig. 5.12 The IA's SC input impedance

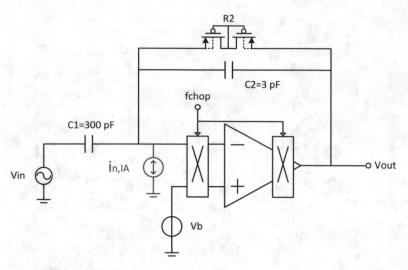

Fig. 5.13 Inverting chopper IA

5.3.2 Chopper Amplifiers with Capacitive Feedback

In other IA architectures, chopper noise will also cause significant excess voltage noise when the input chopper is located at a high-impedance internal node. For example, consider the inverting IA shown in Fig. 5.13, which has been presented in Chap. 3. This IA utilizes a coupling capacitor (C_1) to block the input DC offset, while using a pseudo resistor (R_2) and a capacitor (C_2) in the feedback path to define its voltage gain and establish a high-pass corner at about 0.5 Hz. As a result, the IA's virtual ground is a high-impedance node, which converts chopper noise into a significant amount of excess voltage noise. The IA was implemented in a standard 0.18 μm process, and its input chopper consists of four NMOS devices ($W/L = 0.5/0.18$).

Due to the presence of C_2 in the feedback path, the excess voltage noise PSD exhibits a $1/f^2$ spectrum (with a pole at $1/R_2C_2$), which is given by

$$\overline{i^2_{\text{excess,ICA}}} \approx \overline{i^2_{\text{n,IA}}} \cdot \left(\frac{R_2}{1 + sR_2C_2} \right)^2 \tag{5.11}$$

With chopping disabled, the current noise is quite low, and so any $1/f^2$ noise is buried in $1/f$ noise. This has been verified by periodic noise simulations and measurements (Fig. 5.14), where P_{noise} refers to the periodical noise simulation, and I_{ns} is the noise density of the additional current noise source. As can be seen, the simulation results match the measurement results well.

With chopping enabled, however, the ensuing chopper noise results in an excess of $1/f^2$ voltage noise, which dominates the IA's noise performance (Fig. 5.15). In order to simulate the effect of chopper noise, a current noise source at the

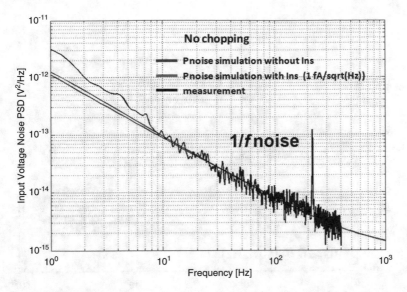

Fig. 5.14 Measured and simulated input-referred noise of an inverting IA without chopping

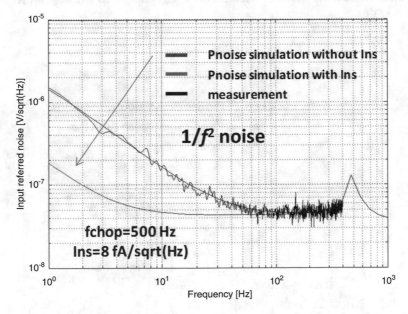

Fig. 5.15 Measured and simulated input-referred voltage noise of an inverting chopper IA at $f_{\text{chop}} = 500$ Hz

high-impedance chopping node (Fig. 5.13) was added. As shown in Figs. 5.15, 5.16, 5.17, and 5.18, its magnitude was then adjusted to fit the measurements obtained at different chopping frequencies. The resulting current noise densities range from

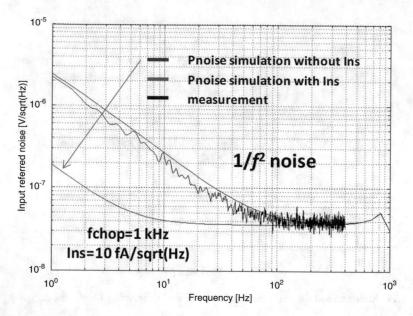

Fig. 5.16 Measured and simulated input-referred voltage noise of an inverting chopper IA at $f_{chop} = 1$ kHz

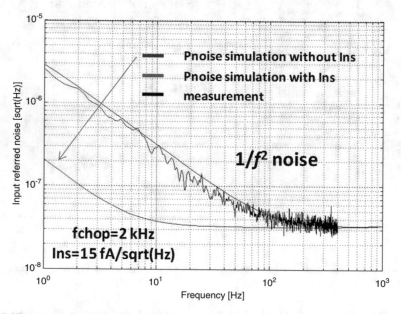

Fig. 5.17 Measured and simulated input-referred voltage noise of an inverting chopper IA at $f_{chop} = 2$ kHz

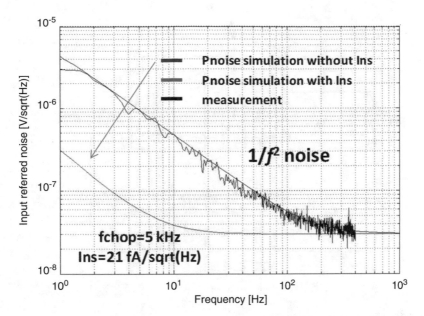

Fig. 5.18 Measured and simulated input-referred voltage noise of an inverting chopper IA at $f_{\text{chop}} = 5$ kHz

7.5 fA/sqrt(Hz) at $f_{\text{chop}} = 500$ Hz (Fig. 5.15) to 21 fA/sqrt(Hz) at $f_{\text{chop}} = 5$ kHz (Fig. 5.18) and are in line with (5.4) and are roughly proportional to the square root of the chopping frequency.

A similar effect occurs in the non-inverting chopper IA [24] shown in Fig. 5.19, which was also implemented in a 0.18 μm process and has been presented in Chap. 4. The IA utilizes a CMOS chopper with equally sized PMOS and NMOS devices ($W/L = 2/0.18$). With chopping disabled, the IA's $1/f$ noise is dominant, and the measured noise is again in good agreement with simulations (Fig. 5.20). With chopping enabled, $1/f^2$ noise becomes dominant since the IA's inverting input is a high-impedance node. The measured noise corresponds to a current noise density from 12 fA/sqrt(Hz) at $f_{\text{chop}} = 500$ Hz to 32.5 fA/sqrt(Hz) at $f_{\text{chop}} = 5$ kHz, as shown in Figs. 5.21, 5.22, 5.23, and 5.24, respectively. Similarly, the current noise density is roughly proportional to the square root of the chopping frequency. In this design, the feedback capacitors were much larger ($16\times$) than those in the inverting IA, and, so, although the $1/f^2$ corner is still dominant, its corner frequency is significantly lower.

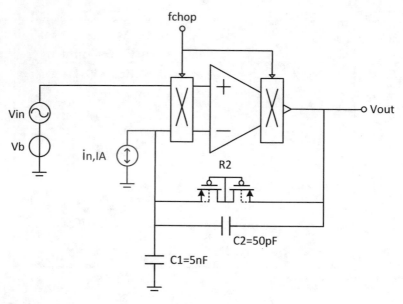

Fig. 5.19 Non-inverting chopper IA

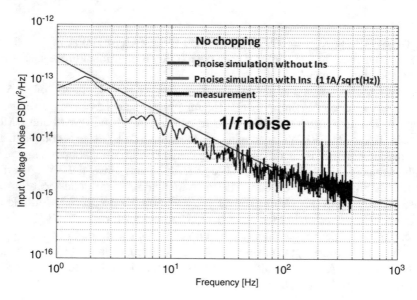

Fig. 5.20 Measured and simulated input-referred voltage noise of a non-inverting IA without chopping

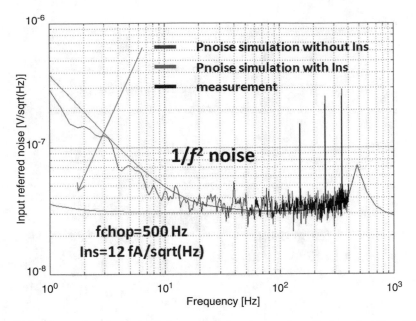

Fig. 5.21 Measured and simulated input-referred voltage noise of a non-inverting IA at $f_{\text{chop}} = 500$ Hz

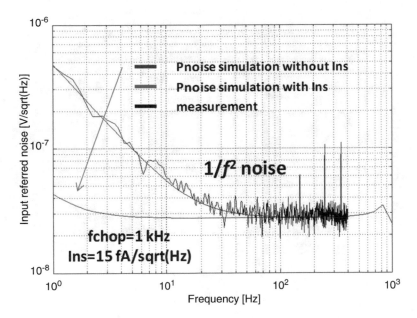

Fig. 5.22 Measured and simulated input-referred voltage noise of a non-inverting IA at $f_{\text{chop}} = 1$ kHz

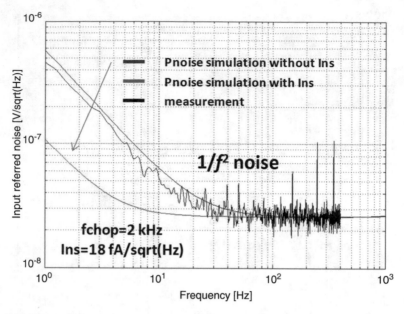

Fig. 5.23 Measured and simulated input-referred voltage noise of a non-inverting IA at $f_{\text{chop}} = 2$ kHz

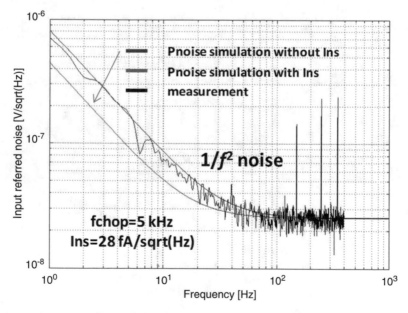

Fig. 5.24 Measured and simulated input-referred voltage noise of a non-inverting IA at $f_{\text{chop}} = 5$ kHz

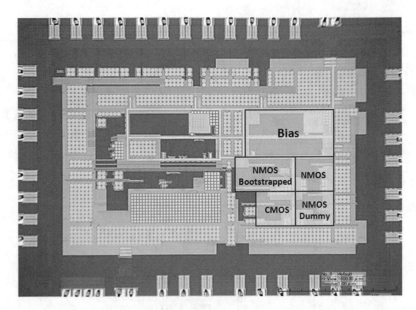

Fig. 5.25 Chip photograph of the noise-testing chip

5.4   A Dedicated Noise-Testing Chip

In order to investigate the relationship between chopper noise, charge injection, and clock feedthrough, a dedicated noise-testing chip was implemented in a 0.18 μm CMOS process (Fig. 5.25). The chip consists of four chopper IAs, similar to the one described in [3], but each equipped with four different types of input choppers (Fig. 5.26): an NMOS chopper, an NMOS chopper with dummy switches, a CMOS chopper, and a bootstrapped NMOS chopper with a low-swing chopper clock. The NMOS chopper was used as a reference, while the other three types of choppers represent various known methods of reducing charge injection and clock feedthrough errors.

As in [25], the bootstrapped NMOS chopper uses a capacitively coupled clock driver to ensure that the MOSFETs are driven at a constant overdrive voltage that is independent of input CM variations. This can also be achieved with a switched-capacitor scheme proposed in [26]. The coupling capacitors and the chopping clock amplitude are chosen such that the amplitude of the resulting V_{gs} is reduced by a factor of 2. To maintain the IA's high input impedance, a voltage follower is used to buffer the input CM voltage and supply the current spikes required by the clock drivers [26].

The current noise PSD of the reference NMOS chopper shows the expected linear relation with the chopping frequency (Fig. 5.27). The current noise PSD produced by the four input choppers is compared in Fig. 5.28. It is interesting to note that both the CMOS chopper and the NMOS chopper with dummy switches generate more current noise than the reference NMOS chopper, while the bootstrapped NMOS

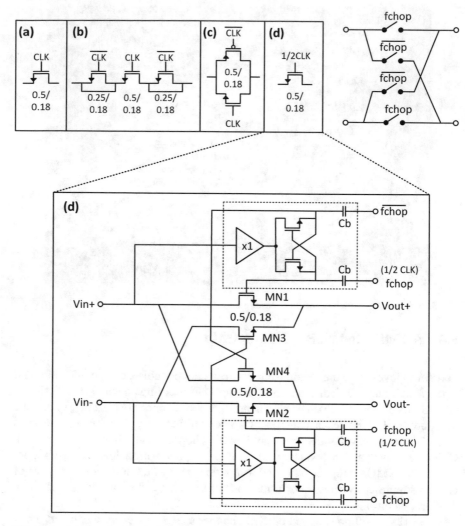

Fig. 5.26 Four types of input chopper switches: (**a**) NMOS, (**b**) NMOS with dummy switches, (**c**) CMOS, (**d**) NMOS with bootstrapped clock drivers

chopper generates the lowest current noise. The reason for this is that chopper noise is related to the charge injection and hence the shot noise, which is associated with the individual chopper switches. As such, it cannot be canceled by using dummy or complementary MOSFETs. In fact, the use of additional MOSFETs only increases the total amount of charge injection and hence the total amount of current noise. However, the bootstrapped NMOS chopper is driven by a low-swing clock, which reduces its charge injection and thus leads to less current noise.

As shown in Fig. 5.29, all the alternative chopper architectures do reduce the IA's DC input current to various degrees. Apparently, the charge injection of the main

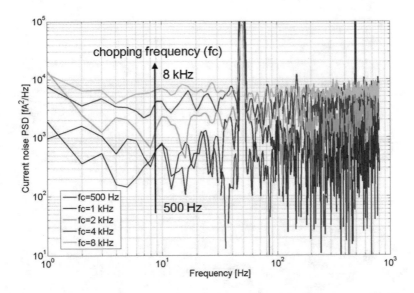

Fig. 5.27 Current noise PSD comparison of the NMOS chopper IA at various chopping frequencies

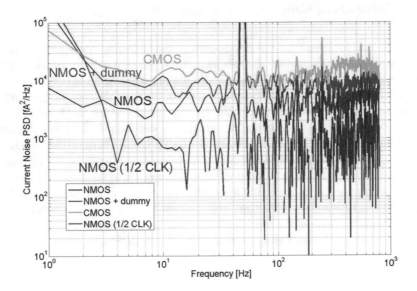

Fig. 5.28 Current noise PSD comparison of four chopper IAs ($f_c = 4$ kHz)

NMOS switches can be significantly reduced by a low-swing clock driver and effectively canceled by the use of simultaneously clocked PMOS or dummy switches, thus leading to lower input currents. As expected from (5.3), the input current of all four chopper IAs increases monotonically with f_{chop}.

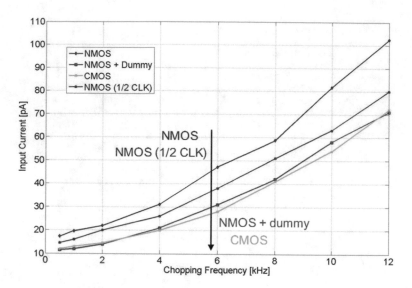

Fig. 5.29 Input current comparison of four chopper IAs

5.5   Methods of Reducing Current Noise

From the measurements on the noise-testing chip, reducing the charge injection associated with the individual chopper switches is the best way to reduce chopper noise. This observation suggests the use of minimum size NMOS or PMOS switches (subject to on-resistance considerations), the lowest possible chopping frequency (subject to $1/f$ noise considerations), and the use of a low-swing clock driver (again subject to on-resistance considerations) to achieve this reduction. In this chapter, a bootstrapped clock driver uses a constant overdrive voltage to drive the MOSFETs of the input chopper. In some cases, it may be possible to simply avoid chopping at high-impedance nodes [27]. In a chopper IA with capacitive feedback, for instance, the input chopper can be located at the output, rather than at the input, of the input stage. This will eliminate the excess voltage noise $1/f^2$ caused by chopper noise. Although the $1/f$ noise of the input stage will now not be chopped, its effect can be somewhat reduced by increasing the size of the input MOSFETs.

5.6   Conclusions

This chapter presents a theoretical analysis and experimental verification of the current noise generated by chopper IAs. This current noise is associated with the charge injection of the input chopper's MOSFET switches, which, in turn, gives rise to a net input current and, we hypothesize, to shot noise. The resulting chopper noise has a white noise spectrum, and its PSD is roughly linearly proportional to the chopping frequency. When chopping is performed at very high-impedance nodes, e.g., in IAs with capacitive feedback networks, chopper noise can cause significant amounts of voltage noise, which may then dominate the amplifier's overall noise performance. The use of a bootstrapped clock driver, which drives the input chopper's MOSFETs with reduced overdrive voltages, is shown to reduce chopper noise.

References

1. C.C. Enz, G.C. Temes, Circuit techniques for reducing the effects of op-amp imperfections: autozeroing, correlated double sampling, and chopper stabilization. Proc. IEEE **84**, 1584–1614 (1996)
2. K.A.A. Makinwa, M.A.P. Pertijs, J.C. van der Meer, J.H. Huijsing, Smart sensor design: the art of compensation and cancellation. *Proc of ESSCIRC*, pp. 76–82 (2007)
3. R.F. Yazicioglu, P. Merken, R. Puers, et al., A 60μW 60nV/√Hz readout front-end for portable biopotential acquisition systems. IEEE J. Solid State Circuits **42**(5), 1100–1110 (2007)
4. G. Ge, C. Zhang, G. Hoogzaad, K.A.A. Makinwa, A Single-Trim CMOS Bandgap Reference with an inaccuracy of ±0.15% from −40 to 125°C. IEEE J. Solid State Circuits **46**(11), 2693–2701 (2011)
5. A. Bakker, K. Thiele, J.H. Huijsing, A CMOS nested-chopper instrumentation amplifier with 100-nV offset. IEEE J. Solid State Circuits **35**(12), 1877–1883 (2000)
6. Q. Fan, J.H. Huijsing, K.A.A.Makinwa, Input characteristics of a chopped multi-path current feedback instrumentation amplifier, *Proc of 4th IEEE IWASI*, (2011)
7. J. Caldwell, Intrinsic noise sources in chopper amplifiers, [online] available: http://e2e.ti.com/cfs-file.ashx/__key/telligent-evolution-components-attachments/00-14-01-00-00-70-21-03/Chopper-Noise.pdf
8. D. Drung, J.-H. Storm, Ultra low-noise chopper amplifier with low input charge injection. IEEE Trans. Instrum. Meas. **60**(7), 2347–2352 (2011)
9. AN3642, Choosing a low-noise amplifier, [online] available: http://pdfserv.maximintegrated.com/en/an/AN3642.pdf
10. K. Blake, Op amp precision design: random noise, Application Note AN1228, Microchip, [online] available: http://ww1.microchip.com/downloads/en/AppNotes/01228a.pdf
11. OPA333 Datasheet, TI, [online] available: http://www.ti.com/lit/ds/symlink/opa333.pdf
12. ADA4051 Datasheet, [online] available: http://www.analog.com/static/imported-files/data_sheets/ADA4051-1_4051-2.pdf
13. ISL28314 Datasheet, Intersil, [online] available: http://www.intersil.com/content/dam/Intersil/documents/fn69/fn6957.pdf
14. J. Xu, Q. Fan, J.H. Huijsing, C. Van Hoof, R.F. Yazicioglu, K.A.A. Makinwa, Measurement and analysis of input current noise in chopper amplifiers. *Proc of ESSCIRC*, 81–84 (2012)
15. J.-H. Shieh, M. Patil, B.J. Sheu, Measurement and analysis of charge injection in MOS analog switches. IEEE J. Solid State Circuits **22**(2), 277–281 (1987)
16. R. Burt, J. Zhang, A micropower chopper-stabilized operational amplifier using a SC notch filter with synchronous integration inside the continuous-time signal path. IEEE J. Solid State Circuits **41**(12), 2729–2736 (2006)
17. A.T.K. Tang, A 3 μV-offset operational amplifier with 20nV/rt(Hz) input noise PSD at DC employing both chopping and autozeroing, *Digest of ISSCC*, pp. 362–387 (2002)
18. K.H. Lundberg, Noise sources in bulk CMOS. [online] available: http://web.mit.edu/klund/www/papers/UNP_noise.pdf

19. R. Sarpeshkar, T. Delbruck, C.A. Mead, White noise in MOS transistors and resistors. IEEE Circuits Devices Magazine **9**(6), 23–29 (1993)
20. J.F. Witte, K.A.A. Makinwa, J.H. Huijsing, The effect of non-idealities in CMOS chopper amplifiers. *Proc of ProRISC*, 616–619 (2004)
21. J.F. Witte, K.A.A. Makinwa, J.H. Huijsing, A CMOS chopper offset-stabilized opamp. IEEE J. Solid-State Circuits **42**(7), 1529–1535 (2007)
22. Switched-Capacitor Network Noise, [online] available: http://inst.eecs.berkeley.edu/~ee247/fa06/lectures/L10_f06_3.pdf
23. D. LaFontaine, Making accurate voltage noise and current noise measurements on operational amplifiers down to 0.1 Hz, Application Note *AN1560*, Intersil, (2011)
24. S. Mitra, J. Xu, A. Matsumoto, K.A.A. Makinwa, C. Van Hoof, R.F. Yazicioglu, A 700µW 8-channel EEG/contact-impedance acquisition system for dry-electrodes. *Digest of Symp. VLSI Circuits*, pp. 68–69 (2012)
25. Q. Fan, J.H. Huijsing, K.A.A. Makinwa, A 78µW ±30V input common-mode range and 160dB CMRR capacitively-coupled chopper instrumentation amplifier with 5µV offset for high-side current-sensing applications. *Digest of ISSCC*, pp. 374–376 (2012)
26. Y. Kusuda, A 5.9nV/√Hz chopper operational amplifier with 0.78µV maximum offset and 28.3nV/°C offset drift. *Digest of ISSCC*, pp. 242–244 (2011)
27. J. Xu, B. Büsze, H. Kim, K.A.A. Makinwa, C. Van Hoof, R.F. Yazicioglu, A 60nV/√Hz 15-channel digital active electrode system for portable biopotential monitoring. *Digest of ISSCC*, 424–425 (2014)

Chapter 6
A Digital Active Electrode System

6.1 IC Architecture Overview

The DAE system (Fig. 6.1) consists of multiple DAE chips and a back-end micro-controller. Each DAE chip (Fig. 6.2) can simultaneously measure biopotential signals (in the ExG channel), real and imaginary ETI signals (in the IMP and IMQ channels), and DC and extremely low-frequency biopotential signals (in the DC channel). The signal chain starts with a chopper IA configured for a voltage gain of 70, which improves the noise/power trade-off of the following programmable gain amplifiers (PGA). The IA contains a ripple reduction loop (RRL) [1] and a DC servo loop (DSL) [2] to attenuate chopper ripple and reject electrode offset, respectively. The IA's output is split into two channels for separate demodulation of the ETI and ExG signals.

The ETI channel is implemented in a similar manner as in the AE described in Chap. 4. Each ETI channel consists of two chopper PGAs that demodulate the ETI signals with inphase f_I (0°) and quadrature-phase f_I (90°) clocks, with respect to a square-wave current injected into the electrode. The DC outputs of IMP and IMQ then represent the real and imaginary components of the ETI, respectively. Compared to the previous implementation, the new PGA of the ExG channel includes a notch filter to reject the ETI signals at f_I (0°). This prevents the PGA's saturation, resulting from the (large) ETI signals that may occur when dry electrodes are used. The detailed operation of the PGA is explained in Sect. 6.2.2.

The DC channel acquires the DC and extreme low-frequency signals present at the output of the DSL. In the frequency domain, the normalized gain and phase of the DC channel complements that of the ExG channel, making it possible to reconstruct DC-coupled ExG signals by combining the outputs of the ExG and DC channels [3, 4].

This chapter is derived from a journal publication by the authors: J. Xu et al., *A 15-Channel Digital Active Electrode System for Multi-Parameter Biopotential Measurement, IEEE J. Solid-State Circuits*, vol.50, no.9, pp. 2090–2100, Sept. 2015.

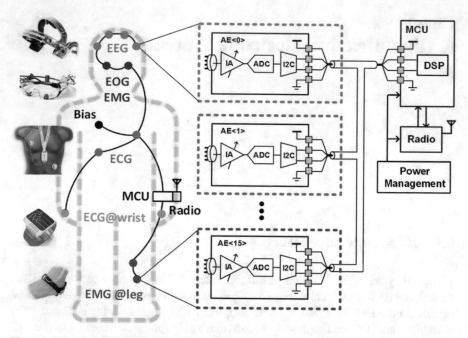

Fig. 6.1 Wearable digital active electrode (DAE) system for multiparameter measurement

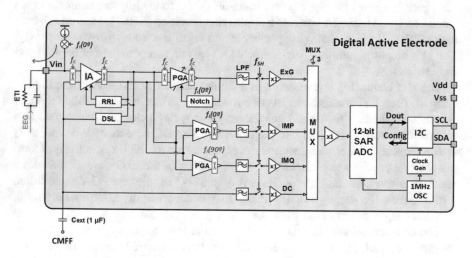

Fig. 6.2 Architecture of a digital active electrode (DAE) chip

These four measurement channels are simultaneously sampled by f_{SH} and connected to an 8-to-1 analog multiplier, through which a back-end microcontroller (µC) selects the channels of interest. A 12-bit SAR ADC [5] digitizes the outputs of these selected channels at 250 S/s to 2 kS/s, and the digital outputs are transmitted to the back-end µC through an I²C interface.

6.2 Analog Signal Processing

6.2.1 A "Functionally" DC-Coupled Instrumentation Amplifier

IAs used in EEG application should tolerate at least ±300 mV electrode offset [6]. For an AE system, each AE should handle the same amount of offset with respective to the subject bias voltage. As discussed in Chap. 2, previous IAs suffered from the trade-off between electrode offset tolerance, input impedance, noise, and power [7–9]. To tolerate a large electrode offset, the gain of a truly DC-coupled amplifier will be reduced by the need to avoid clipping [7]. Therefore, a high-resolution ADC will be required to digitize the small (μV) biopotential signals superimposed on a much larger (mV) electrode offset, leading to high power dissipation on each channel. In contrast, AC-coupled amplifiers, implemented with passive coupling capacitors, enable low-power rail-to-rail offset rejection [2, 8, 9]. But this comes at the cost of filtering out the DC and low-frequency signals. Capacitively coupled chopper IAs mitigate $1/f$ noise by chopping the input signal before the coupling capacitor [10, 11], but their input impedance is limited by switched-capacitor resistance. AC-coupled IAs with voltage-to-current DSLs solve this problem [3, 12]. However, these IAs only compensate for a few tens of millivolts of electrode offset, limited either by noise considerations or by the maximum current provided from the feedback loop.

This section presents a "functionally" DC-coupled chopper IA architecture with a voltage-to-voltage feedback (Fig. 6.3) to facilitate large electrode offset tolerance while still optimizing the IA's performance trade-off. The core IA is implemented with a current feedback IA [12], chopped at 4 kHz to reduce its flicker noise. This IA architecture provides high input impedance (100 MΩ at 50 Hz) and wide input CM range (0.5–1.2 V), making it robust to electrode impedance mismatch and DC polarization from dry electrodes. The DSL consists of a gm-C integrator that monitors the output offset and then cancels it by driving the core IA's inverting input. As a result, the IA can reject up to ±350 mV of electrode offset, which is determined by the amplifier's input CM range and noise specification.

An interesting feature of this IA architecture, as well as of any IA equipped with a DSL configured in voltage-to-voltage feedback, is the preservation of the DC and low-frequency information, which is available at the output of the DSL. The IA's normalized AC and DC outputs have complementary gain and phase characteristics, so these two outputs can be combined to implement a "functionally" DC-coupled IA. It has the same transfer function as a truly DC-coupled IA (Fig. 6.4) but with a much wider DC dynamic range (±350 mV) in conjunction with a high AC gain (>40 dB). The wide DC dynamic range mitigates electrode offset from dry electrodes, while the high AC gain relaxes the required ADC resolution.

The DSL utilizes a weak transconductance ($g_{m2} = 3$ μS) [13] and an external capacitor ($C_{ext} = 1$ μF) to achieve a low cutoff frequency (<0.5 Hz). In addition, a large C_{ext} reduces the impedance at the input chopping node, reducing the $1/f^2$ noise

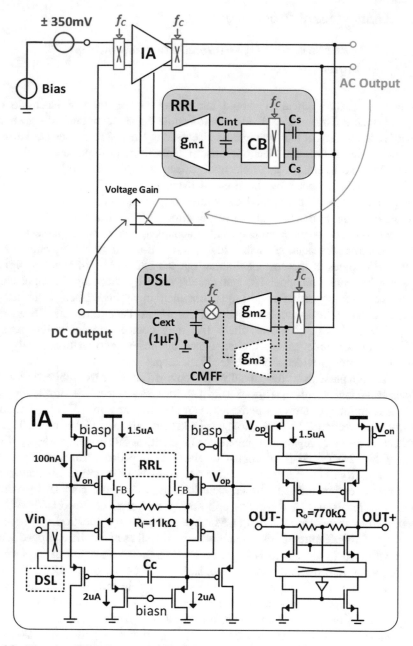

Fig 6.3 "Functionally" DC-coupled chopper IA

generated by the IA's input current noise [14]. To suppress the $1/f$ noise of the DSL and the RRL, both loops are chopped. The thermal noise PSD of the core IA, the RRL and the DSL, respectively, is given by

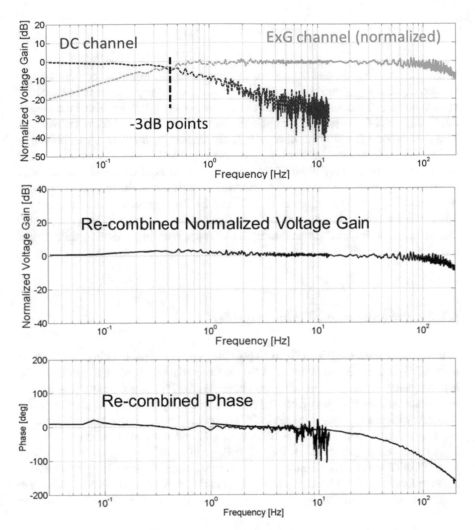

Fig. 6.4 Measured normalized gain of ExG and DC channel (by dividing the respective channel gain measured at analog outputs) and the recombined gain and phase.

$$\overline{V_{n,\text{IA}}^2} = \frac{\overline{V_{n,\text{coreIA}}^2}}{\left(1 + \frac{C_s g_{\text{m1}} R_o}{C_{\text{int}}}\right)^2} \tag{6.1}$$

$$\overline{V_{n,\text{RRL}}^2} = \left(\overline{V_{n,\text{gm1}}^2} + \frac{\overline{I_{n,\text{CB}}^2}}{s^2 C_{\text{int}}^2}\right) \cdot g_{\text{m1}}^2 R_{\text{i}}^2 \tag{6.2}$$

$$\overline{V_{n,\text{DSL}}^2} = \overline{V_{n,\text{gm2}}^2} \cdot \frac{g_{\text{gm2}}^2}{s^2 C_{\text{ext}}^2} \tag{6.3}$$

where $V_{n,\text{coreIA}}$, $V_{n,\text{gm1}}$, and $V_{n,\text{gm2}}$ are the input-referred noise of the core IA and transconductance g_{m1} and g_{m2}, respectively, $I_{n,\text{CB}}$ is the input current noise of the current buffer, R_i and R_o are the input and output resistors that determine the gain of the core IA, and C_s is the RRL's input capacitance for voltage-to-current conversion. The total input-referred noise of an AE is dominated by the core IA as shown in (6.1). Large integrator capacitors ($C_{\text{int}} = 150$ pF and $C_{\text{ext}} = 1$ μF) are selected to minimize the noise contribution of the core IA as well as of the RRL and the DSL.

At start-up, the circuit takes tens of seconds to settle due to the large time constant of the weak g_{m2} and the external C_{ext} (Fig. 6.5). To overcome this issue, the AE includes a foreground fast-settling path, so that a stronger g_{m3} ($=100 g_{\text{m2}}$) in parallel can be temporarily switched on during the start-up, ensuring a settling time of less than 1 s.

6.2.2   Programmable Gain Amplifier

In the ExG channel, the PGA (Fig. 6.6) provides a programmable gain that facilitates both EEG and ECG applications. Chopper modulation is used to mitigate the low-frequency $1/f$ noise. A notch filter attenuates the ETI signal before it is filtered by the succeeding low-pass filter (LPF). The operating principle is similar to the RRL. Any inphase ETI signal at the PGA's output is first converted to an AC current via C_s, which is then demodulated back to DC and integrated on capacitor C_{int}. Transconductor g_{m4} up-converts the DC voltage to an AC current and feeds it back to the PGA (Fig. 6.7). This feedback current compensates for the ETI current flowing through R_i. On the other hand, the ExG signal at the PGA's output is up-modulated to 1 kHz and so is suppressed by C_{int}. The PGA can attenuate the output ETI signal by

$$A_{V,\text{PGA@1kHz}} = \frac{G_{\text{PGA}}}{\frac{R_{\text{out,CB}} \cdot g_{\text{m4}} \cdot R_{\text{o,PGA}}}{Z_{\text{s@1kHz}}} + 1} \approx \frac{Z_{\text{s@1kHz}}}{R_{\text{out,CB}} \cdot g_{\text{m4}} \cdot R_{\text{i,PGA}}} \tag{6.4}$$

where $Z_s = C_s // R_{\text{in,CB}}$, $R_{\text{in,CB}}$, and $R_{\text{out,CB}}$ are the input and output DC resistance of the current buffer (CB), respectively. $R_{\text{i,PGA}}$ and $R_{\text{o,PGA}}$ are the PGA's internal feedback resistors. To maximize the attenuation, the PGA utilizes a cascode current buffer and a large input resistor ($R_{\text{i,PGA}} = 1$ MΩ). Figure 6.8 shows that the notch filter can reduce the output ETI signal ($f_{\text{ETI}} = 1$ kHz) by a factor of 40.

The core PGA utilizes the same IA architecture but with single-ended output. The coarse gain (2, 10, and 20) is selected via $R_{\text{o,PGA}}$, while R_{DAC} is implemented with a 12-bit programmable resistor array and can be trimmed with 50 Ω resolution. To achieve this goal, very large CMOS switches ($W/L = 500/0.18$) are used. R_{DAC} can be used to trim the channel gain of two DAEs and so can improve the CMRR at the

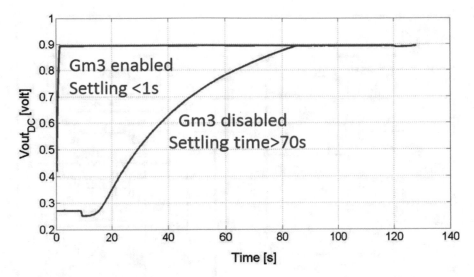

Fig. 6.5 Measured settling time with and without g_{m3}

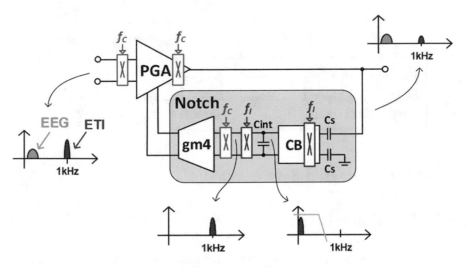

Fig. 6.6 PGA with a notch filter for ETI signal rejection

analog outputs by about 5 dB. However, the CMRR improvement at the digital outputs is obscured by the 12-bit ADC's quantization [15]. Instead of trimming, a common-mode feedforward (CMFF) technique can also improve the CMRR of two DAEs. This will be discussed in detail in Sect. 6.4.

The IMP and IMQ channels also include PGAs for a wide range of the ETI measurements. The PGA does not have a notch filter and only contains an output chopper for ETI demodulation, because an ExG signal typically has a lower magnitude than an ETI input signal.

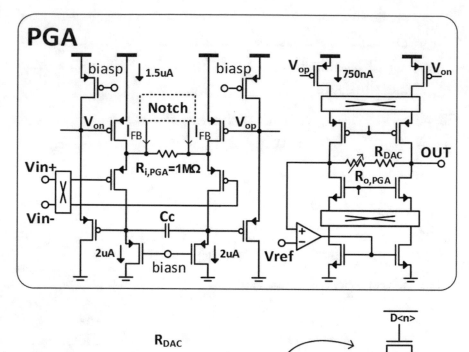

Fig. 6.7 Core PGA with a programmable resistor load

6.3   Digital Interfaces

The built-in digital interface is responsible for data transmission between the AEs and the digital back end (DBE) as well as for clock signal generation of the AEs. An I^{2}C interface is selected because it only requires two wires (SCL and SDA) for bi-directional communication (Fig. 6.9), and it is compatible with many commercially available μCs. Although the equally popular SPI interface can operate at higher clock speeds (up to tens of MHz), it requires four wires (MISO, MOSI, clock, and a separate chip selection to each IC), which would substantially increase the system's wiring bulk.

Compared to a standard I^{2}C interface, the proposed digital interface allows a global read and writes to all DAE sensors. The global write configures and synchronizes DAEs at each I^{2}C cycle. The global read enables DAEs to sequentially transfer

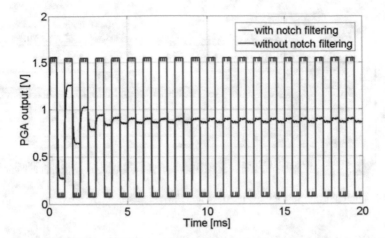

Fig. 6.8 Simulated ETI signal at PGA's output ($G_{PGA} = 5$, $V_{ETL\_input} = 280$ mV$_{pp}$ at 1 kHz)

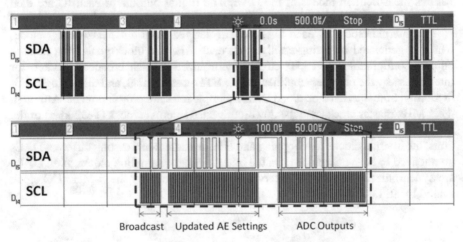

Broadcast   Updated AE Settings   ADC Outputs

Fig. 6.9 Data (SDA) and clock (SCL) signals of an AE's I^{2}C bus

the data back to the master node with only a single command. This avoids the need for the I^{2}C master to address each DAE individually, thus reducing the control overhead and the amount of data toggling on the bus.

Each individual DAE chip can be given a 4-bit address via 4 external pins, allowing up to 16 DAEs to be connected to a single μC. To align the sampling moments of the individual DAE nodes, the back-end μC first sends a broadcast packet to all ICs (Fig. 6.10). This broadcast packet (I^{2}C address = 0) is identified by each DAE chip independent of its base address on the I^{2}C bus. The broadcast packets align the sample-and-hold (f_{SH}) and ADC sampling clocks of each DAE and also select two internal measurement channels of each IC (via MUX < 1:0 > in Fig. 6.2), whose outputs will be sent to the μC in the next I^{2}C cycle. In this way, the back-end

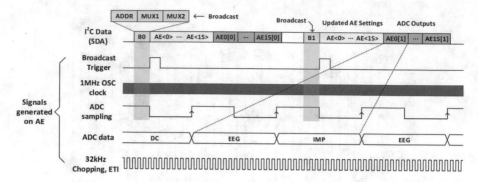

Fig. 6.10 Timing diagram of the I^{2}C interface and internal clocks

µC has full control of the DAE and can flexibly select any channel of interest. The broadcast packet is followed by configuration settings from the µC, including various measurement modes of the DAE. The digital outputs of each IC are then transmitted to the µC during the next I^{2}C cycle.

The internal clocks of each DAE IC are derived from a 1 MHz master clock, which is generated by a ring-oscillator on each AE. The clock generation module outputs configurable 1–32 kHz clock synchronized at each broadcast packet, for internal use by the chopper amplifier, ADC, ETI measurement, and digital logic. For flexibility, both internal clocks and DAE's sample rates are programmable. Although the 1 MHz master clocks and the down-converted internal clocks (1–32 kHz) of the DAEs suffer from frequency variations (~10%) due to the oscillators' PVT variations, the internal clocks for chopping and ETI measurements among different DAEs do not need to be synchronized. On the other hand, the sampling clocks of all DAEs are synchronized at every I^{2}C interval (1 ms in default) by the 5 MHz I^{2}C clock shared by all DAEs.

6.4 CMRR Enhancement

There are two different AE mismatch mechanisms that limit the CMRR between a pair of AEs. The first is an electrode impedance mismatch, which is actually more of a problem with dry electrodes. In this case, the AE should maintain very large input impedance over the entire ExG bandwidth to mitigate any voltage division. The second is the AEs' gain mismatch. Compensating for these mismatches can significantly improve the CMRR and signal quality.

A further challenge of the DAE system is the need to achieve a wide CM input dynamic range for each AE. This is because each DAE can be modeled as a single-ended amplifier with a large gain (up to 1400), followed by a 12-bit ADC. As a result, any CM aggressor (e.g., mains interference, motion artifacts) that appears at the DAE's input can easily distort or saturate its readout circuits, even in the absence of any gain mismatch between the DAEs.

Previous designs employed feedback techniques to improve CMRR and input CM dynamic range. The driven-right-leg (DRL) approach [16], for instance, has been widely used to compensate for CM interference by feeding the CM signal back to the subject through a bias electrode (or ground electrode). However, the DRL may suffer from instability because the loop gain is not well defined, especially with the large and ill-defined electrode impedance associated with the use of dry electrodes. Common-mode feedback (CMFB) circuits can solve this problem by feeding the CM signal back to the amplifier's input, instead of the subject. However, an analog CMFB circuit [8] relies heavily on large passive components and results in poor flexibility and area efficiency. Alternatively, a digitally-assisted CMFB scheme [15] extracts the CM signal of all DAEs in the digital domain, converts it back to an analog signal via a DAC, and then feeds it back to each DAE. However, the latency induced by the I^2C bus significantly shifts the phase of the analog CMFB signal relative to the input CM signal. This results in reduced phase margin and may destabilize the CMFB loop. To mitigate this stability problem, the bandwidth and gain of the CMFB loop has to be sacrificed [15]. Another major issue with all these "feedback-based" techniques is their instability during the electrode "lead off" condition. Since the common-mode extraction loop is broken, any electrode making poor electrical contact can cause the failure of the system [17].

The system utilizes a new and more generic CMFF technique to improve the CMRR of two AEs, providing advantages over a previous CMFB technique [15] in terms of higher CMRR bandwidth, better power efficiency, and stability. Furthermore, this CMFF technique is generic and applicable to different types of AE architectures, such as inverting amplifiers or non-inverting amplifiers, whereas the CMFF technique proposed in [18] is only suitable for non-inverting amplifiers. The key idea of the new CMFF technique (Fig. 6.11) is to apply an input CM signal to all the DAEs before pre-amplification. The input CM signal is applied to the inverting inputs of all IAs via a buffer and their capacitively coupled DSLs. Therefore, the input CM signal to a pair of AEs is compensated at their differential outputs, although their differential ExG amplification is not affected. This new CMFF technique has another major advantage: the buffered CM signal is applied to all DAEs through a very low impedance, which reduces the noise and interference pickup from the environment, similar to the noise reduction principle of an active electrode.

In the detailed implementation (Fig. 6.11), the input CM signal can be acquired from an additional electrode, or simply from the reference electrode, or from any one of the recording electrodes. This flexible selection is based on the fact that the DAEs' input CM signals picked up from the environment are quite similar. In extreme cases where the electrodes are placed far from each other, several local CMFF schemes can be used for different groups of DAEs.

Figure 6.12 illustrates the setup of CMRR measurement with various electrode impedance mismatch scenarios. Figures 6.13, 6.14, and 6.15 show the measured differential-mode gain and common-mode gain of a pair of DAEs, with different

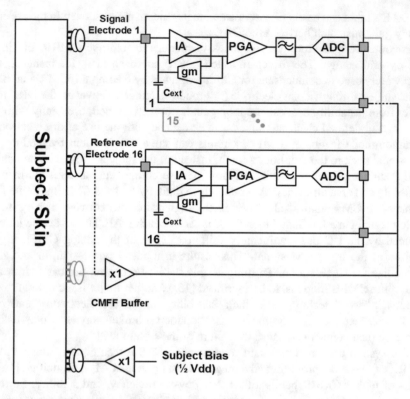

Fig. 6.11 CMRR improvement through the use of a CMFF electrode

resistors R_e (0 Ω, 50 kΩ, and 800 kΩ) to mimic different electrode types and their impedance mismatch [19]. The CMFF technique significantly boosts the CMRR of an AE pair from 40 to 102 dB (at 50 Hz). When R_e increases, the CMFF is less effective due to the attenuation of the input CM signal and the larger gain mismatch between AEs.

6.5   Measurement

6.5.1   Measurement of Performance

The DAE was implemented in a standard 1P6M 0.18 µm CMOS process and occupies an area of 15.8 mm^2 (Fig. 6.16). Each chip consumes 58 µA from a 1.8 V core supply, excluding the I^{2}C interface.

Each ExG channel (consisting of two DAEs) shows a 60 nV/sqrt(Hz) input-referred noise density (Fig. 6.17), which stays constant over ±350 mV electrode offset with respect to the subject bias (Fig. 6.18). Each DAE has an input current

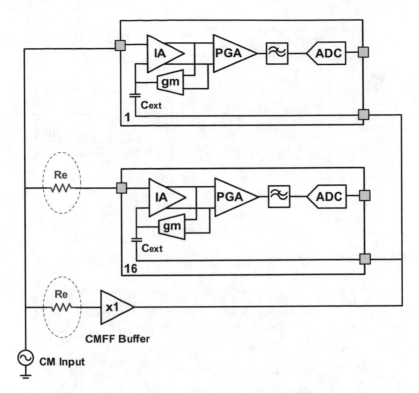

Fig. 6.12 CMRR measurement at various electrode impedance conditions

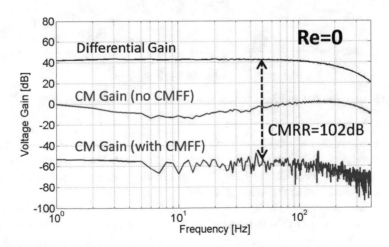

Fig. 6.13 Measured DM gain and CM gain versus frequency ($R_e = 0$)

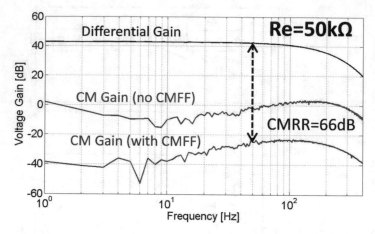

Fig. 6.14 Measured DM gain and CM gain versus frequency ($R_e = 50$ kΩ)

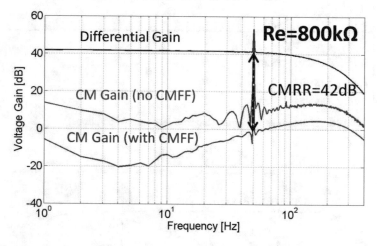

Fig. 6.15 Measured DM gain and CM gain versus frequency ($R_e = 800$ kΩ)

noise density of 20 fA/sqrt(Hz) at a chopping frequency of 4 kHz (Fig. 6.19) and an input impedance of 100 MΩ at 50 Hz (Fig. 6.20). Each DAE can measure up to 400 kΩ resistance (Fig. 6.21) at 1 kHz (at gain of 140) when measured by connecting multiple test resistors to the input of a DAE.

Table 6.1 summarizes the IC's performance. The analog performance is competitive to that of state-of-the-art biopotential IAs. The proposed IA achieves a good balance between noise, electrode offset tolerance, and CMRR. Furthermore, the major merits of the proposed DAE system are the AE-based architecture for low interference, a built-in digital interface for high integration, and an inter-chip CMRR

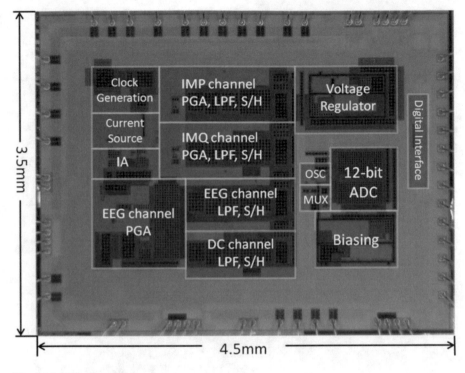

Fig. 6.16 Chip photograph

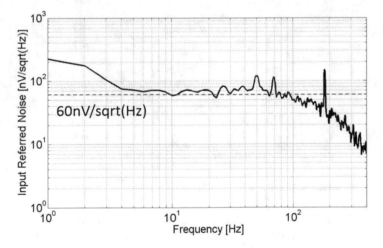

Fig. 6.17 Measured input-referred noise per ExG channel ($G = 700$)

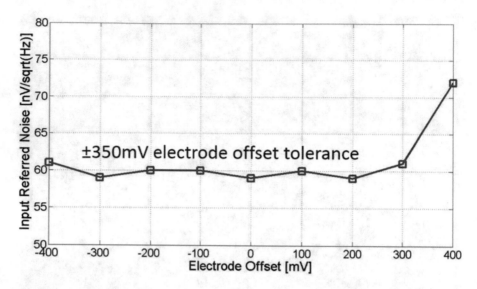

Fig. 6.18 Measured input noise per ExG channel versus electrode offset

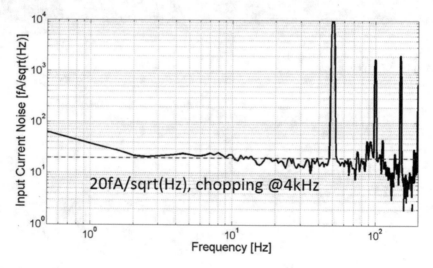

Fig. 6.19 Measured input current noise of a digital AE

boosting technique. These features eliminate the need for an additional analog back-end (BE) circuit, leading to a cost-efficient solution for multichannel ExG acquisition.

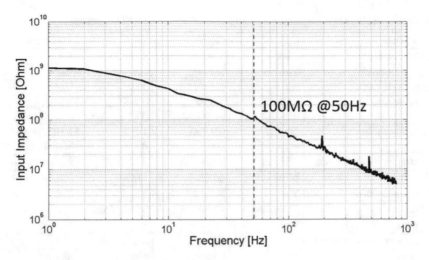

Fig. 6.20 Measured input impedance of a DAE

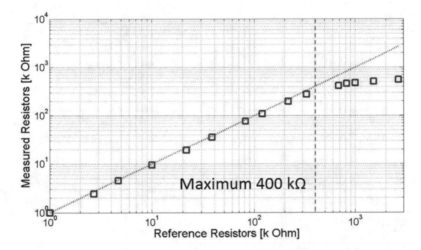

Fig. 6.21 Measured resistor values versus reference resistor values, showing the dynamic range of the ETI readout

6.5.2   Multiparameter ExG Measurement

Simultaneous single-channel ECG, EMG, and EOG measurements are performed to demonstrate the DAEs' capability of multiparameter acquisition. Five (wet) electrodes are attached to a subject's chest and forehead (Fig. 6.1) and connected to DAE test boards via cables. These five electrodes include one bias electrode, two electrodes for ECG recording, and two electrodes for EMG and EOG recording. For simplicity, the CMFF buffer's input is connected to the reference electrodes in all

Table 6.1 Performance summary compared to the state-of-the-art EEG systems

Parameters	[18]	[2]	[7]	This work
Supply voltage	1.8 V	1 V	2.7–5.25 V	1.8 V
Active electrode	Yes	No	No	Yes
DC-coupled IA	No	No	Yes	Yes
Input-referred noise	1.75μV$_{rms}$ (0.5–100 Hz)	1.3μV$_{rms}$ (0.5–100 Hz)	0.7μV$_{rms}$ (DC-131 Hz)	0.65μV$_{rms}$ (0.5-100 Hz)
Input impedance	1.2 GΩ @20 Hz	0.7 GΩ (DC)	1 GΩ (DC)	1 GΩ@1 Hz, 300 MΩ@20 Hz
Electrode offset tolerance	±250 mV	Rail-to-Rail	±250 mV	±350 mV
CMRR	84 dB	60 dB	115 dB	102 dB
ETI measurement	Yes	No	Yes	Yes
ADC	12-bit SAR	12-bit SAR	24-bit SDM	12-bit SAR
Number of channels	8	18	8	15
Dry electrode applications	Yes	No	No	Yes
Current (per channel)	48 μA	>3.5 μA	250 μA	58 μA

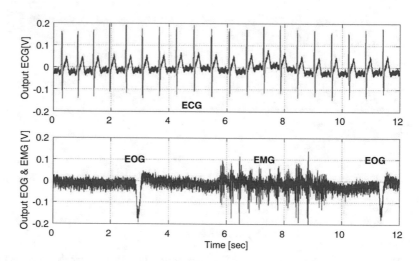

Fig. 6.22 Simultaneous ExG recordings of the DAE system

measurements. Figure 6.22 shows several types of physiological behaviors acquired by the DAE system, such as heartbeat (ECG), face muscle movement (EMG), and eyes blinking (EOG).

In order to measure EEG on the scalp, five DAE test boards (each board contains a DAE ASIC, level shifters, analog test buffers, and jumpers for I^{2}C address) are connected in a daisy chain and attached to an EEG headset (Fig. 6.23). The bias electrodes, reference electrodes, and (signal) recording electrodes are placed at O_1, O_2, C_z, P_z, C_3, and C_4, respectively, based on a standard 10–20 electrodes EEG system. Figure 6.24 shows that alpha activity at approximately 12 Hz clearly allows

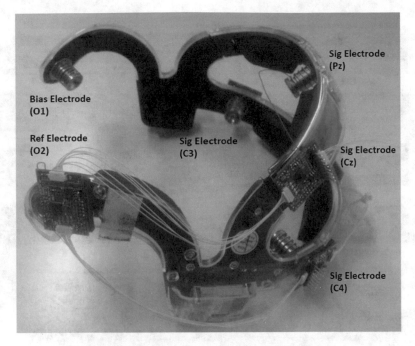

Fig. 6.23 4-channel EEG headset with DAE test boards attached

distinguishing between periods of "eyes open" and "eyes closed" when dry electrodes are used.

Figure 6.25 shows a 1-lead ECG measured on the subject's wrists, demonstrating the benefit of the CMFF: after enabling the CMFF scheme, the 50 Hz interference, picked up from the same environment, is significantly reduced.

Figure 6.26 shows a simultaneous recording of the ECG and ETI (on the chest). When one electrode is disconnected from the subject, the ECG output shows incorrect results and the ETI output saturates. After reconnecting the electrode, both the ECG and ETI recover from saturation. This indicates that the ETI output can also be used for instant lead-on and lead-off detection.

6.6 Conclusions

A digital active electrode (DAE) ASIC incorporates amplifiers, an ADC, and a digital interface on a single chip. A "functionally" DC-coupled IA optimizes performance trade-offs (between noise, electrode offset tolerance, input impedance, and power) and enables the practical use of dry electrodes. A generic CMFF technique ensures the maximum 102 dB CMRR of two DAEs at 50 Hz. The highly integrated DAE chips eliminate the needs for a back-end analog signal processor and facilitate

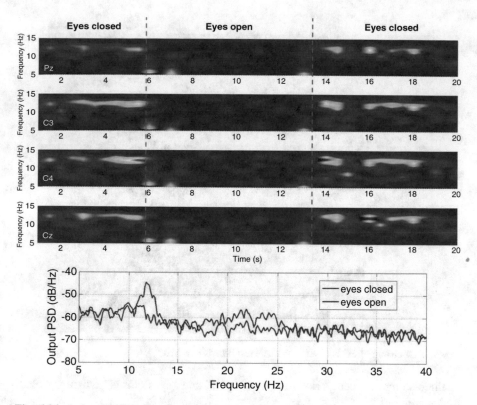

Fig. 6.24 4-channel EEG recording with dry electrodes during periods of eyes closed and eyes open

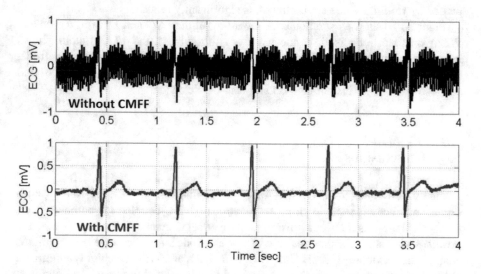

Fig. 6.25 1-lead ECG recording with wet electrodes placed on wrists

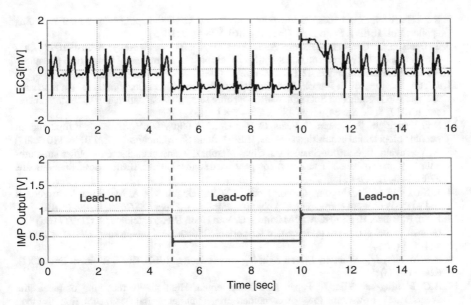

Fig. 6.26 Lead-off and lead-on detection by monitoring the ETI output

multichannel multiparameter biopotential signal acquisition. These highly modular and scalable DAEs significantly reduce the system's complexity and cost.

References

1. R. Wu, K.A.A. Makinwa, J.H. Huijsing, A chopper current-feedback instrumentation amplifier with a 1mHz 1/f noise corner and an AC-coupled ripple reduction loop. IEEE J. Solid State Circuits **44**(12), 3232–3243 (2009)
2. N. Verma, A. Shoeb, A.J. Bohorquez, J. Dawson, J. Guttag, A.P. Chandrakasan, A micro-power EEG acquisition SoC with integrated feature extraction processor for a chronic seizure detection system. IEEE J. Solid State Circuits **45**(4), 804–816 (2010)
3. R. Muller, S. Gambini, J.M. Rabaey, A 0.013mm^2 2.5µW, DC-coupled neural signal acquisition IC with 0.5V supply. IEEE J. Solid State Circuits **47**(1), 232–243 (2012)
4. P. Schoenle, F. Schulthess, R. Ulrich, F. Huang, T. Burger, Q. Huang, A DC-connectable multichannel biomedical data acquisition ASIC with mains frequency cancellation. *Proc. ESSCIRC*, 149–152 (2013)
5. R.F. Yazicioglu, K. Sunyoung, T. Torfs, H. Kim, C. Van Hoof, A 30µW analog signal processor ASIC for portable biopotential signal monitoring. IEEE J. Solid-State Circuits **46**(1), 209–223 (2011)
6. IEC60601-2-26, Medical electrical equipment – Part 2–26: Particular requirements for the basic safety and essential performance of electroencephalographs
7. TI-ADS1298, 8-channel, 24-bit analog-to-digital converter with integrated ECG front end. Texas Instruments, [online] available: <http://www.ti.com/lit/ds/symlink/ads1298.pdf>
8. J. Xu, R.F. Yazicioglu, P. Harpe, K.A.A. Makinwa, C. Van Hoof, A 160µW 8-channel active electrode system for EEG monitoring. *Digest of ISSCC*, 300–302 (2011)

9. R.R. Harrison, C. Charles, A low-power low-noise CMOS amplifier for neural recording applications. IEEE J. Solid State Circuits **38**(6), 958–965 (2003)
10. T. Denison, K. Consoer, W. Santa, A.-T. Avestruz, J. Cooley, A. Kelly, A 2μW 100nV/√Hz chopper-stabilized instrumentation amplifier for chronic measurement of neural field potentials. IEEE J. Solid State Circuits **42**(12), 2934–2945 (2007)
11. J. Yoo, Y. Long, D. El-Damak, M.A.B. Altaf, A.H. Shoeb, A.P. Chandrakasan, An 8-channel scalable EEG acquisition SoC with patient-specific seizure classification and recording processor. IEEE J. Solid State Circuits **48**(1), 214–228 (2013)
12. R.F. Yazicioglu, P. Merken, R. Puers, C. Van Hoof, A 60μW 60nV/√Hz readout front-end for portable biopotential acquisition systems. IEEE J. Solid State Circuits **42**(5), 1100–1110 (2007)
13. A. Veeravalli, E. Sanchez-Sinencio, J. Silva-Martinez, Transconductance amplifier structures with very small transconductances: A comparative design approach. IEEE J. Solid State Circuits **37**(6), 770–775 (2002)
14. J. Xu, Q. Fan, J.H. Huijsing, C. Van Hoof, R.F. Yazicioglu, K.A.A. Makinwa, Measurement and analysis of input current noise in chopper amplifiers. *Proc. ESSCIRC*, 81–84 (2012)
15. J. Xu, B. Büsze, H. Kim, K.A.A. Makinwa, C. Van Hoof, R.F. Yazicioglu, A 60nV/√ (Hz) 15--channel digital active electrode system for portable biopotential monitoring. *Digest of ISSCC*, 424–425 (2014)
16. B.B. Winter, J.G. Webster, Driven-right-leg circuit design. IEEE Trans. Biomed. Eng. **30**(1), 62–66 (1983)
17. A.C. Metting-van Rijn, A. Peper, C.A. Grimbergen, High-quality recording of bioelectric events. Part 2. Low-noise, low-power multichannel amplifier design. Med. Biol. Eng. Comput. **29**(4), 433–440 (1991)
18. J. Xu, S. Mitra, A. Matsumoto, S. Patki, C. Van Hoof, K.A.A. Makinwa, R.F. Yazicioglu, A wearable 8-channel active-electrode EEG/ETI acquisition system for body area networks. IEEE J. Solid State Circuits **49**(9), 2005–2016 (2014)
19. Y.M. Chi, T.-P. Jung, G. Cauwenberghs, Dry-contact and noncontact biopotential electrodes: methodological review. IEEE Rev. Biomed. Eng. **3**, 106–119 (2010)

Chapter 7
Conclusions

7.1 Summary

The prototype ASICs described in Chaps. 3, 4, and 6 demonstrate that active electrodes (AEs) can be successfully used for wearable EEG acquisition. The highly integrated and ultralow-power AEs are compatible with dry electrodes, thus facilitating different form factors for wearable devices. From a user point of view, these user-friendly features are their principal advantages over conventional wet-electrode-based, bulky, and power-hungry EEG instruments.

Apart from the improved user comfort, the proposed AE systems also achieve state-of-the-art performance through the use of advances in IC techniques. Various ultralow-power IC design techniques have been implemented and verified in different types of AEs, whose analog performance is compared in Table 7.1. The combination of chopping and capacitive feedback IA architecture helps the AE achieve low-noise amplification and rail-to-rail electrode offset rejection (Chap. 3). An AE's input impedance can be further improved through the use of an impedance boosting loop (Chap. 3) or a non-inverting IA topology (Chap. 4). To compensate for the AEs' mismatch, either a CMFB technique (Chap. 3 and 6) or a CMFF technique (Chap. 4 and 6) was implemented, improving the CMRR by at least 25 dB. The non-idealities of a chopper IA, such as intrinsic offset and chopper ripple, can be reduced by digitally assisted calibration techniques (Chap. 3) or by a continuous-time ripple reduction loop (RRL) (Chap. 4 and 6). Furthermore, the AE systems described also introduce the electrode-tissue impedance (ETI) measurement (Chap. 4 and 6) and a "functionally" DC-coupled IA (Chap. 6), both of which aim to provide additional information of the brain-electrode interface beyond EEG recording. In general, the DAE presented in Chap. 6 achieves the best overall performance while also including the most features. Compared to this, state-of-the-art AE implementations either consume significant power [1, 2], which requires main power supplies, or have less analog performance, power efficiency, or functionalities [3, 4].

© Springer International Publishing AG 2018
J. Xu et al., *Low Power Active Electrode ICs for Wearable EEG Acquisition*, Analog Circuits and Signal Processing, https://doi.org/10.1007/978-3-319-74863-4_7

Table 7.1 Performance summary of the AE systems presented in Chap. 3, 4, and 6

	Chapter 3	Chapter 4	Chapter 6
Technology/supply	0.18 μm/ 1.8 V	0.18 μm/ 1.8 V	0.18 μm/1.8 V
AE gain	3, 10, 100	11, 51, 101	140, 700, 1200
Input-referred noise (per channel)	1.2μV$_{rms}$ (0.5–100 Hz)	1.75μV$_{rms}$ (0.5–100 Hz)	0.65μV$_{rms}$ (0.5–100 Hz)
Electrode offset rejection	Rail-to-rail	±250 mV	±300 mV
Input impedance at DC and 50 Hz	2 GΩ 120 MΩ	1.2 GΩ 400 MΩ	1 GΩ 100 MΩ
CMRR at 50 Hz	82 dB (via CMFB)	84 dB (via CMFF)	102 dB (via CMFF)
Power consumption (per channel)	20 μW + g.tec (N/A)	82 μW	105 μW (excl. digital interface)
Integrated ADC	N/A	12-bit SAR	12-bit SAR
Ripple reduction	Foreground	Background	Background
ETI measurement	No	Yes	Yes
DC coupling	AC coupling	AC coupling	"Functionally" DC coupling
Integrated digital interface	No	No	I²C

In spite of these advantages, AE systems still have some drawbacks or limitations. For example, each EEG recording channel consists of two AEs, inherently resulting in a lower power efficiency than conventional differential EEG amplifiers. Furthermore, chopping at a high-impedance node may also generate significant $1/f^2$ noise because of the current noise of the chopper switches. Although general design guidelines have been discussed to mitigate this effect (Chap. 5), the $1/f^2$ noise was not completely eliminated. Lastly, although a DC servo loop using voltage-to-voltage feedback (Chap. 6) represents an excellent balance between noise, input impedance, and electrode offset rejection, it requires the use of a large off-chip capacitor and is not suitable for low-supply voltages (<0.6 V).

7.2   Future Work

In general, there are three major research objectives for the future development of wearable EEG ICs and brain monitoring systems: better suppression of motion artifacts, improved robustness and safety, and multimodal acquisition.

Dealing with motion artifacts is one remaining challenge for improving signal quality. The dynamic range of an EEG readout circuit is typically limited to a few mVs because of the IA's gain constraints. As a result, the IA can saturate during the presence of large motion artifacts, especially when the subject is moving. This can be a severe problem for wearable devices extensively used in lifestyle and wellness applications. One straightforward solution is to reduce the IA's gain; however, a high-resolution ADC (>16 bit) with low-power consumption would be needed. Another possible solution is to apply a motion artifact reduction (MAR) technique [5]. With this technique, the motion artifact signal can be partially extracted from electrode-tissue impedance (ETI) measurements and can be used to compensate the input motion artifact. However, the accuracy of the MAR not only depends on a high-quality ETI measurement but also on digital signal processing to ensure that the original EEG signal is not polluted through the MAR.

Improving the robustness and safety of the existing wearable EEG systems in special medical environments is another interesting objective. One example of this is the use of an EEG headset during functional magnetic resonance imaging (fMRI). This simultaneous EEG-fMRI recording is a multimodal neuroimaging technique, which enables the measurements of both neuronal and hemodynamic activities. However, the fMRI environment can lead to particular problems for EEG acquisition. For instance, large currents induced by the fMRI acquisition process may flow into and thus saturate the EEG device.

Emerging brain monitoring systems should also support additional physiological modalities for improved diagnostic accuracy. For instance, measuring the blood oxygenation response through an optical sensor or estimating the fluid status and body composition through a bio-impedance measurement. These measurements can be combined with EEG recordings to examine the brain's functional activities more comprehensively [6]. Recently, multiparameter biopotential signal acquisition systems [7, 8] containing multiple types of sensors have been presented. These systems can be easily attached to people's heads, arms, chests, or wrists for simultaneous measurement, from which various biopotential signals (ECG, ETI, bio-impedance, or fNIRS) are recorded and wirelessly transmitted to medical professionals through a body area network (BAN).

References

1. ActiveTwo. [online] available: http://www.biosemi.com/activetwo_full_specs.htm
2. g.USBamp. [online] available: http://www.gtec.at/Products/Hardware-and-Accessories/g. USBamp-Specs-Features
3. M. Guermandi, R. Cardu, E. Franchi, R. Guerrieri, Active electrode IC combining EEG, electrical impedance tomography, continuous contact impedance measurement and power supply on a single wire. *Digest of ESSCIRC*, pp. 335–338 (2011)
4. Y.M. Chi, C. Maier, G. Cauwenberghs, Ultra-high input impedance, low noise integrated amplifier for noncontact biopotential sensing. IEEE J. Emerg. Selected Top. Circuits Syst. **1** (4), 526–535 (2011)
5. N. Van Helleputte, S. Kim, H. Kim, J.P. Kim, C. Van Hoof, R.F. Yazicioglu, A 160µW biopotential acquisition IC with fully integrated IA and motion artifact suppression. IEEE Trans. Biomed Circuits Syst. **6**(6), 552–561 (2012)
6. P. Castellone, Combining EEG with NIRS. Brain Products. [online] available: http://www. brainproducts.com/files/public/products/brochures_material/pr_articles/1304_EEG-NIRS.pdf

7. N. Van Helleputte et al., A 345µW multi-sensor biomedical SoC with bio-impedance, 3-channel ECG, motion artifact reduction, and integrated DSP. IEEE J. Solid State Circuits **50**(1), 230–244 (2015)
8. U. Ha et al., A wearable EEG-HEG-HRV multimodal system with real-time tES monitoring for mental health management. *Digest of ISSCC*, pp. 1–3 (2015)

Summary

This book describes the application, theory, and implementation of active electrodes (AEs) for EEG acquisition systems that require low noise, high input impedance, high electrode offset tolerance, high CMRR, and low power. The motivation for selecting AEs is to enable the use of high-impedance dry electrodes with improved robustness to environmental interference and cable motion. In turn, dry electrodes facilitate long-term EEG measurements with greater user comfort. Three generations of AE-based ASICs were implemented with different architectures and circuit design techniques.

Chapter 1 introduces the basics of scalp EEG measurement, the history of its development, and the need for personal EEG devices. AEs are shown to be a promising solution for dry-electrode-based EEG measurement, and the associated design challenges are summarized.

Chapter 2 presents an overview of state-of-the-art instrumentation amplifiers (IAs) and AEs for wearable healthcare. Different architectures and design techniques are presented and compared, which aim to optimize key specifications such as noise level, input impedance, electrode offset tolerance, CMRR, and power dissipation.

Chapter 3 presents an AE readout circuit based on an AC-coupled chopper amplifier, which naturally blocks electrode offset. The use of chopping mitigates $1/f$ noise, resulting in an input-referred noise of 0.8 μV_{rms} (0.5–100Hz). An impedance-boosting technique increases its input impedance by $5\times$ (at 1 Hz), while digitally assisted offset trimming reduces residual ripple and offset by $20\times$ and $14\times$, respectively. Mismatch between the AEs is the main cause of a low CMRR. To mitigate this, a back-end common-mode feedback (CMFB) circuit improves the CMRR of a pair of AEs by 30 dB.

Chapter 4 presents a complete eight-channel AE system for continuous monitoring of EEG and electrode-tissue impedance (ETI). ETI measurement extends system functionality by enabling remote assessment of electrode status. The whole AE system consists of nine AEs and one back-end (BE) analog signal processor (ASP). The AE is based on a non-inverting chopper amplifier, which improves EEG recording with a good trade-off between input impedance and noise level.

The BE circuit post-processes and digitizes the AEs' analog outputs. At the system level, a common-mode feed-feedback (CMFB) technique improves the CMRR of an AE pair by 25 dB.

Chapter 5 investigates the root cause of $1/f^2$ noise of chopper amplifiers through a theoretical analysis and measurements of several chopper IAs. We hypothesize that the charge injection and clock feedthrough associated with the MOSFETs of the input chopper give rise to significant input current and current noise. In combination with high source impedances, this "chopper noise" is converted to voltage noise, which may then be a significant contributor (i.e., $1/f^2$ noise) to the IA's total input-referred voltage noise. Furthermore, the chopper noise has a white power spectral density, whose magnitude is roughly proportional to the chopping frequency. Design guidelines are then proposed to reduce the chopper noise. A further proposal is the use of a clock-bootstrapped chopper, which exhibits less noise than a traditional chopper.

Chapter 6 presents a digital active electrode (DAE) system for multiparameter biopotential signal acquisition. It is built around an ASIC that performs analog signal processing and digitization through on-chip instrumentation amplifiers, a 12-bit ADC, and a digital interface. Via a standard I^2C bus, up to 16 DAEs (15 channels) can be connected to a microcontroller, thus significantly simplifying the system's connection. At the circuit level, a DAE uses a "functionally" DC-coupled amplifier to handle extremely low-frequency biopotential signals while still tolerating high levels of electrode offset. At the system level, a more generic common-mode feedforward (CMFF) technique improves the CMRR of an AE pair from 40 dB to the maximum of 102 dB.

Chapter 7 concludes the book by comparing the overall performance of the AEs presented in Chaps. 3, 4, and 6, illustrating both their advantages and limitations with respect to conventional EEG acquisition ICs. Three research tracks, namely, suppression of motion artifacts, improved robustness and safety, as well as multi-modal acquisition, are proposed for future work.

List of Publications

Journal Papers

J. Xu, R.F. Yazicioglu, B. Grundlehner, P. Harpe, K.A.A. Makinwa, C. Van Hoof, A 160μW 8-channel active electrode system for EEG monitoring. IEEE Trans. Biomed. Circuits Syst. **5**(6), 555–567 (2011)

J. Xu, Q. Fan, J.H. Huijsing, C. Van Hoof, R.F. Yazicioglu, K.A.A. Makinwa, Measurement and analysis of current noise in chopper amplifiers. IEEE J. Solid State Circuits **48**(7), 1575–1584 (2013)

J. Xu, S. Mitra, A. Matsumoto, S. Patki, K.A.A. Makinwa, C. Van Hoof, R.F. Yazicioglu, A wearable 8-channel activeelectrode EEG/ETI acquisition system for body area networks. IEEE J. Solid State Circuits **49**(8), 2005–2016 (2014)

J. Xu, B. Büsze, C. Van Hoof, K.A.A. Makinwa, R.F. Yazicioglu, A 15-channel digital active electrode system for multiparameter biopotential measurement. IEEE J. Solid State Circuits **50**(9), 2090–2100 (2015)

J. Xu, S. Mitra, C. Van Hoof, R.F. Yazicioglu, K.A.A. Makinwa, Active electrodes for wearable EEG acquisition: review and electronics design methodology. IEEE Rev. Biomed. Eng. **10**, 187–198 (2017)

Conference Papers

S. Mitra, J. Xu, A. Matsumoto, K.A.A. Makinwa, C. Van Hoof, R.F. Yazicioglu, A 700μW 8-channel EEG/contact-impedance acquisition system for dry-electrodes. *Digest of Symp. VLSI Circuits*, pp. 68–69 (2012)

J. Xu, R.F. Yazicioglu, P. Harpe, K.A.A. Makinwa, C. Van Hoof, A 160μW 8-channel active electrode system for EEG monitoring. *Digest of ISSCC*, pp. 300–302 (2011)

J. Xu, Q. Fan, J.H. Huijsing, C. Van Hoof, R.F. Yazicioglu, K.A.A. Makinwa, Measurement and analysis of current noise in chopper amplifiers. *Digest of ESSCIRC*, pp. 81–84 (2012)

J. Xu, B. Büsze, H. Kim, K.A.A. Makinwa, C. Van Hoof, R.F. Yazicioglu, A 60nV/√(Hz) 15-channel digital active electrode system for portable biopotential monitoring. Digest of ISSCC, 424–425 (2014)

© Springer International Publishing AG 2018
J. Xu et al., *Low Power Active Electrode ICs for Wearable EEG Acquisition*, Analog Circuits and Signal Processing, https://doi.org/10.1007/978-3-319-74863-4

J. Xu, P. Harpe, J. Pettine, C. Van Hoof, R.F. Yazicioglu, A low power configurable bio-impedance spectroscopy ASIC with simultaneous ECG and respiration recording functionality. *Digest of ESSCIRC*, pp. 396–399 (2015)

J. Xu et al., A 36μW reconfigurable analog front-end IC for multimodal vital signs monitoring, *Digest of Symp. VLSI Circuits*, pp.170–171 (2017)

Book Chapters

N. Van Helleputte, J. Xu, et al., Advances in biomedical sensor systems for wearable health, in *Hybrid ADCs, Smart Sensors for the IoT, and Sub-1V & Advanced Node Analog Circuit Design*, (Springer International Publishing, Cham, 2018), pp. 121–143

J. Xu, R. Mohan, N. Van Helleputte, S. Mitra, Design and optimization of ICs for wearable EEG sensors, in *CMOS Circuits for Biological Sensing and Processing*, (Springer International Publishing, Cham, 2018), pp. 163–185

R.F. Yazicioglu, J. Xu, et al., Low power biomedical interfaces, in *Efficient Sensor Interfaces, Advanced Amplifiers and Low Power RF Systems*, (Springer International Publishing, Cham, 2016), pp. 81–101

Index

A
AC-coupled inverting amplifiers, 14, 16
Active electrode (AEs), 5–8, 116
 See also Back-end common-mode
  feedback (CMFB) circuit
 back-end summing amplifier, 23
 bio-amplifiers techniques, 23
 biopotential EEG measurement, 40–45
 cable motion and interference, 38–43
 EEG readout circuit, 23, 24
 IA (*see* Instrumentation amplifier (IA))
 IC measurement, 36–40
 input impedance boosting, 27–31 (*see also*
  Noise analysis)
 RRL and DSL, 24
 single-ended IAs, 23
 See also Backnd commonode feedback
  (CMFB) circuit
Ambulatory EEG systems, 1
Analog signal processing, functionally
 DC-coupled IA, 95–98

B
Back-end (BE) analog signal process
 EEG and ETI channels, 57
 level shifters, 58
 LPFs and ADC, 59
 TC and TI stages, 58
Back-end common-mode feedback (CMFB)
 circuit
 diagram of, 33, 34
 DRL and, 35, 36
 dry electrodes, 36
 equivalent circuits, 32, 34

 reference inputs, 33
Battery-powered AE system, 8
Bio-amplifier architectures
 AC-coupled inverting amplifiers, 14
 analog buffers, 14
 instrumentation amplifiers, 17
Bio-amplifier design techniques
 chopper modulation, 11
 DRL technique, 13
Bio-impedance measurement, 117
Biopotential EEG measurement, 40–45
Body area network (BAN), 117
Brain-computer interfaces (BCI), 59
Brain monitoring systems, 116

C
Cable motion and interference, 38–43
Charge injection, 70–72
Chopper amplifiers
 capacitive feedback, 80–86
 conventional CMOS-/JFET-input
  amplifiers, 69 (*see also* Current noise
  analysis)
 description, 69
 MOSFETs implementation, 69
 precision signal conditioning, 69
Chopper modulation technique, 11, 12, 23, 24
Clock driver circuit, 72, 73
Clock feedthrough, 70–72
Common-mode (CM) signal extraction, 23
Common-mode feedback (CMFB) circuit, 9
Common-mode feedforward (CMFF)
 technique, 9, 115
 actual DC voltage, 53

Common-mode feedforward (CMFF) technique
    (*cont.*)
    AEs' reference inputs, 51
    CM averaging, 53
    conventional AEs without CMFF, 52
    DRL circuit, 51
Common-mode rejection ratio (CMRR), 8, 9,
    11, 13, 102–106
Current-balancing instrumentation, 18
Current noise analysis
    charge injection and clock feedthrough,
        70–72
    clock driver circuit, 72, 73
    CMOS amplifier, 70
    conventional Chopper modulated IA, 75–79
    equivalent circuit model, 70
    parasitic SC resistance, 73, 74
    reduction methods, 90
    shot noise, 72
    typical MOSFET switch, 74

D
DC-coupled amplifier, 16, 95
DC servo loop (DSL), 13, 16, 17, 23, 24, 26–29,
    31, 54
Digital active electrode (DAE) system, 9
    IC architecture, 93, 94
    multiparameter measurement, 94
    performance measurement, 104–110
Digital interfaces
    ADC sampling, 101
    AEs and DBEs, 100
    clock generation module, 102
    I^2C interface and internal clocks, 102
    SDA and SCL signals, 101
    standard I^2C interface, 100
Digitally-assisted offset compensation, 19
Driven-right-leg (DRL) approach, 13, 14, 103
Dry electrodes, 5

E
Eight-channel AE system, 59–65
Electrical domain, 3
Electrode offset rejection, 12–19, 55
Electrode-tissue impedance (ETI), 7, 13, 57,
    58, 117
    BE signal chain, 49
    EEG measurement, 50, 51
    eight-channel AE-based system, 49, 50
    eight-channel EEG/ETI acquisition
        system, 50
    PGA and LPF, 49

Electrode-tissue interface, 3, 4
Electroencephalograms (EEGs), 1, 116
Emerging brain monitoring systems, 117

F
$1/f^2$ noise, 80, 83
Four-channel wireless EEG headset, 65, 66
Fourth-order Sallen-Key LPFs, 59
Functionally DC-coupled amplifiers, 18–19
Functionally DC-coupled IA, 19, 95–98
Functional magnetic resonance imaging
    (fMRI), 117

I
Impedance boosting technique, 12, 23, 24,
    27–31
Impedance bootstrapping, 11–12
Input impedance boosting, 27–31
Instrumentation amplifier (IA), 6, 17, 18
    AE, 54
    chopper modulation, 24
    current steering DACs, 24, 25
    digitally assisted ripple and offset reduction,
        26–29
    feedback capacitors ratios, 24
    input impedance, 54
"Interference generator", 40
Interference reduction test, 42

L
Low-pass filters (LPF), 11, 49, 59

M
Motion artifact reduction (MAR)
    technique, 117
Multiparameter ExG measurement, 109–113

N
Noise analysis
    chopping frequency, 32
    core amplifier, AE, 57
    feedback capacitors, 55
    $1/f^2$ voltage noise source, 56
    input equivalent circuit, 31, 32
    input-referred noise specification, 55
    NMOS and PMOS differential pairs, 56
    non-inverting topology, 55, 56
    pseudo-resistor, 31
    shaping factors, 31, 33

Noise efficiency factor (NEF), 15
Noise-testing chip, 87–89
Non-inverting AC-coupled amplifiers, 16–17

O
Offset compensation, 12–13

P
Parasitic switched-capacitor (SC) resistance,
    73, 74
Portable EEG devices, 1
Programmable gain amplifier (PGA)
    CMRR improvement, 99
    EEG and ECG applications, 98
    IMP and IMQ channels, 99
    inphase ETI signal, 98

    with programmable resistor load, 100
    simulated ETI signal, 101
Programmable gain amplifiers (PGA), 49
PWM communication, 53–54

R
Ripple reduction loop (RRL), 11, 23, 24,
    26–29, 31, 54, 55

S
Shot noise analysis, 72
Skin-electrode interface, 6

W
Wearable EEG measurement, 2

Printed in the United States
By Bookmasters